Contents

Content Guidance

Questions & Answers

Getting the most from this book

Examiner tips

Advice from the examiner on key points in the text to help you learn and recall unit content, avoid pitfalls, and polish your exam technique in order to boost your grade.

Knowledge check

Rapid-fire questions throughout the Content Guidance section to check your understanding.

Knowledge check answers

Ι Turn to the back of the book for the Knowledge check answers.

Summary

Summaries

● Each core topic is rounded off by a bullet-list summary for quick-check reference of what you need to know.

Questions & Answers

Exam-style questions

Examiner comments on the questions
Tips on what you need to do to gain full marks, indicated by the icon ℯ.

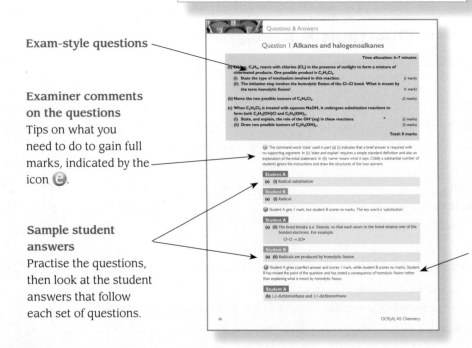

Examiner commentary on sample student answers
Find out how many marks each answer would be awarded in the exam and then read the examiner comments (preceded by the icon ℯ) following each student answer.

Sample student answers
Practise the questions, then look at the student answers that follow each set of questions.

About this book

This unit guide is the second in a series of two, which together cover the OCR AS chemistry specification A. This guide is written to help you to prepare for Unit Test 2, which examines the content of **Unit F322: Chains, Energy and Resources**.

The **Content Guidance** section gives a point-by-point description of all the facts you need to know and concepts that you need to understand for Unit F322. It aims to provide you with a basis for your revision. However, you must also be prepared to use other sources in your preparation for the examination. The Content Guidance section gives special emphasis to the reagents, conditions, balanced equations and mechanism for each reaction.

The **Question and Answer** section shows you the sort of questions you can expect in the unit test. It would be impossible to give examples of every kind of question in one book, but the questions used should give you a flavour of what to expect. Each question has been attempted by two candidates, student A and student B. Their answers, along with the examiner's comments, should help you to see what you need to do to score a good mark — and how you can easily *not* score marks, even though you probably understand the chemistry.

What can I assume about the guide?

You can assume that:
- the topics covered in the Content Guidance section relate directly to those in the specification
- the basic facts you need to know are stated clearly
- the major concepts you need to understand are explained
- the questions at the end of the guide are similar in style to those that will appear in the unit test
- the answers supplied are genuine, combining responses commonly written by students
- the standard of the marking is broadly equivalent to the standard that will be applied to your answers

What can I *not* assume about the guide?

You must *not* assume that:
- every last detail has been covered
- the way in which the concepts are explained is the *only* way in which they can be presented in an examination (often concepts are presented in an unfamiliar situation)
- the range of question types presented is exhaustive (examiners are always thinking of new ways to test a topic)

Study skills and revision techniques

All students need to develop good study skills. This section provides advice and guidance on how to study AS chemistry.

Organising your notes

Chemistry students often accumulate a large quantity of notes, so it is useful to keep these in a well-ordered and logical manner. It is necessary to review your notes regularly, maybe rewriting the notes taken during lessons so that they are clear and concise, with key points highlighted. You should check your notes using textbooks, and fill in any gaps. Make sure that you go back and ask your teacher if you are unsure about anything, especially if you find conflicting information in your class notes and textbook.

It is a good idea to file your notes in specification order using a consistent series of headings. The Content Guidance section can help you with this.

Organising your time

Preparation for examinations is a very personal thing. Different people prepare, equally successfully, in very different ways. The key is being totally honest about what actually *works for you.*

Whatever your style, you must have a plan. Sitting down the night before the examination with a file full of notes and a textbook does not constitute a revision plan — it is just desperation — and you must not expect a great deal from it. Whatever your personal style, there are a number of things you *must* do and a number of other things you *could* do.

Unit F322 contains lots of new terms and concepts. There is also a large body of factual knowledge, relating to reagents and reaction conditions, which have to be learnt. Constant repetition is particularly helpful, so throughout this guide you are given exercises to help improve your memory and understanding.

The scheme outlined below is a suggestion as to how you might revise Unit F322 over a 3- to 4-week period.

Day	Week 1	Week 2
Mon	Naming, isomerism, calculations and alkanes — allow about 30 min	There are five major sections to the organic chemistry — you should now have five sheets of A4 as your summary notes Reread the summary notes — 15 min Get someone to test you on them — 15 min
Tue	Alkenes — allow about 20–25 min Reread yesterday's notes — 5 min	Reread summary notes — 10 min Write them out again from memory — 20 min
Wed	Fuels — allow about 20 min Reread yesterday's notes — 5 min Go over Monday's section — 3–4 min	Reread summary notes on alkanes — 2 min, and try an exam question on alkanes — 15 min Reread summary notes on alkenes — 2 min, and try an exam question on alkenes — 15 min
Thu	Alcohols — allow about 20 min Reread yesterday's notes — 5 min Go over Tuesday's section — 3–4 min Go over Monday's section — 1 min	Reread summary notes on fuels — 2 min, and try an exam question on fuels — 15 min Reread summary notes on alcohols — 2 min, and try an exam question on alcohols — 15 min
Fri	Halogenoalkanes and modern analytical techniques (MAT) — allow about 20 min Reread yesterday's notes — 5 min Go over Wednesday's section — 3–4 min Go over Tuesday's section — 1 min	Reread summary notes on halogenoalkanes — 2 min, and try an exam question on halogenoalkanes — 15 min

OCR(A) AS Chemistry

Day	Week 1	Week 2
Sat	Reread notes on halogenoalkanes and MAT — allow 5–10 min Go over Thursday's section — 3–4 min Summarise alkanes on one side of A4 paper Summarise alkenes on one side of A4 paper	Use past papers and try four or five questions from organic chemistry — allow 30 min. Mark your answers and make a list of anything you do not understand. Ask your teacher for help with things you do not understand or are unsure of
Sun	Reread notes on halogenoalkanes and MAT — allow 5 min Summarise fuels on one side of A4 paper Summarise alcohols on one side of A4 paper Summarise halogenoalkanes on one side of A4 paper	Use past papers and try four or five questions from organic chemistry — allow 30 min Mark your answers and make a list of anything you do not understand. Ask your teacher for help with things you do not understand or are unsure of

Day	Week 3	Week 4
Mon	Enthalpy changes — allow about 30 min	Reread the summary notes — 10 min Write them out again from memory — 20 min
Tue	Rates — allow about 20–25 min Reread yesterday's notes — 5 mins	Reread the summary notes on enthalpy changes — 2 min, and try an exam question on enthalpy changes — 15 min Reread summary notes on rates — 2 min, and try an exam question on rates — 15 min
Wed	Equilibrium — allow about 20 min Reread yesterday's notes — 5 min Go over Monday's section — 3–4 min	Reread summary notes on equilibrium — 2 min, and try an exam question on equilibrium — 15 min Reread summary notes on resources — 2 min, and try an exam question on resources — 15 min
Thu	Resources — allow about 20 min Reread yesterday's notes — 5 min Go over Tuesday's section — 3–4 min Go over Monday's section — 1 min	You should now have nine summary sheets which cover the entire syllabus. Reread them all and write them out again from memory — 30 min
Fri	Reread yesterday's notes — 5 min Go over Wednesday's section — 3–4 min Go over Tuesday's section — 1 min Summarise enthalpy changes on one side of A4 paper	Use past papers and try four or five questions — allow 30 min. Mark your answers and make a list of anything you do not understand. Ask your teacher for help with things you do not understand or are unsure of
Sat	Go over Thursday's section — 3 min Go over Wednesday's section — 2 min Summarise rates on one side of A4 paper Summarise equilibrium on one side of A4 paper Summarise resources on one side of A4 paper	Use past papers and try four or five questions — allow 30 min. Mark your answers and make a list of anything you do not understand. Ask your teacher for help with things you do not understand or are unsure of
Sun	You should now have four sides of A4 as your summary notes for this section of the syllabus Reread summary notes — 15 min Get someone to test you on them — 15 min	Use past papers and try four or five questions — allow 30 min. Mark your answers and make a list of anything you do not understand. Ask your teacher for help with things you don't understand or are unsure of

This revision timetable may not suit you, in which case write one to meet your needs. It is only there to give you an idea of how one might work. The most important thing is that the grid at least enables you to see what you should be doing and when you should be doing it. Do not try to be too ambitious — *little and often is by far the best way.*

It would of course be sensible to put together a longer rolling programme to cover all your AS subjects. Do *not* leave it too late. Start sooner rather than later.

Content Guidance

Module 1: Basic concepts and hydrocarbons

Nomenclature and formulae

Nomenclature should follow IUPAC rules for naming organic compounds. The IUPAC (International Union of Pure and Applied Chemistry) rules for naming are based around the systematic names given to the alkanes and the prefix or suffix given to each functional group.

In this module you will be expected to recognise and to name alkanes, alkenes, alcohols and halogenoalkanes.
- Alkanes have a general formula of C_nH_{2n+2}.
- Alkenes have a general formula of C_nH_{2n}.
- Alcohols have a general formula of $C_nH_{2n+1}OH$.
- Halogenoalkanes have a general formula of $C_nH_{2n+1}X$, where X = F, Cl, Br or I.

It is important that you learn the appropriate prefix or suffix for each group:
- Alkanes always end in '-ane'.
- Alkenes always end in '-ene'.
- Alcohols always end in '-ol'.
- Halogenoalkanes always start with 'fluoro-', 'chloro-', 'bromo-' or 'iodo-'.

Compounds A, B and C are named in Figure 1.

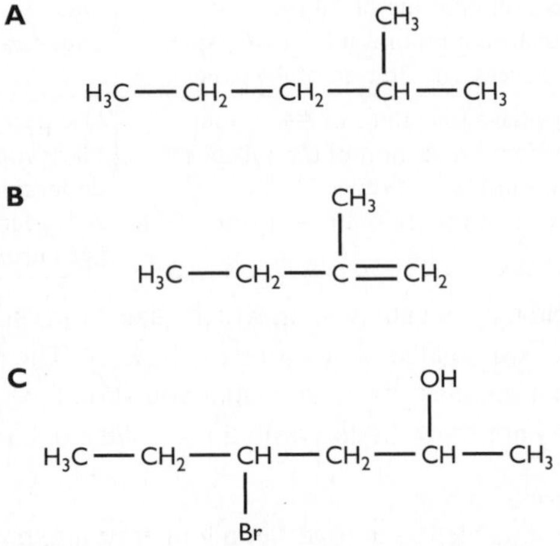

A

$$H_3C — CH_2 — CH_2 — \overset{\displaystyle CH_3}{\overset{|}{CH}} — CH_3$$

B

$$H_3C — CH_2 — \overset{\displaystyle CH_3}{\overset{|}{C}} = CH_2$$

C

$$H_3C — CH_2 — \underset{\underset{\textstyle Br}{|}}{CH} — CH_2 — \overset{\displaystyle OH}{\overset{|}{CH}} — CH_3$$

Figure 1

OCR(A) AS Chemistry

- **A** is an alkane and hence ends in ...*ane*. The longest carbon chain is five, hence *pent*ane. There is a branch (on the second carbon atom) that consists of one carbon, hence *2-methyl*, giving the full name *2-methylpentane*.
- **B** is an alkene and hence ends in ...*ene*. The longest carbon chain is four, hence *but*ene. The double bond starts at carbon atom one, hence *but-1-ene*. There is a branch (on the second carbon atom) that consists of one carbon, hence *2-methyl*, giving the full name of *2-methylbut-1-ene*.
- **C** is both an alcohol (and hence ends in ...*ol*) and a bromoalkane (and hence starts with *bromo*...). The longest carbon chain is six, hence *hex*ane. The alcohol is on the second carbon (hence *-2-ol*) and the bromine is on the fourth carbon (hence *4-bromo*...). The full name is *4-bromohexan-2-ol*.

You will be expected to draw and to use structural formulae, displayed formulae and skeletal formulae.

A **structural formula** is accepted as the minimal detail, using conventional groups, for an unambiguous structure. The molecular formula C_4H_{10} could be either of two isomers, butane or methylpropane, and is therefore ambiguous. The structural formula for butane is $CH_3CH_2CH_2CH_3$, and the structural formula of methylpropane is written as $(CH_3)_3CH$.

A **displayed formula** shows both the relative placing of atoms and the number of bonds between them. The displayed formulae for butane and methylpropane are shown in Figure 2.

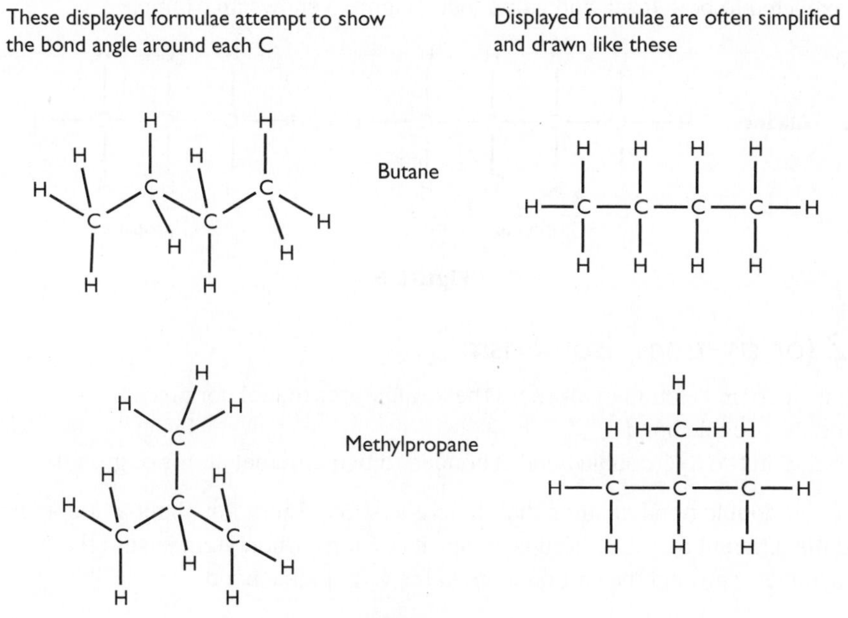

Knowledge check 1

Name the following:
$CH_3CHBrCH_3$,
$CH_3CHCHCH_3$, $(CH_3)_4C$,
$CH_3CH_2CH(CH_3)CH_2CH_3$.

These displayed formulae attempt to show the bond angle around each C

Displayed formulae are often simplified and drawn like these

Butane

Methylpropane

Figure 2

Knowledge check 2

Draw the displayed formula for each of 2-chloropropane, butan-2-ol and 3-methylbut-1-ene.

A **skeletal formula** is used to show a simplified organic formula by removing hydrogen atoms from alkyl chains, leaving only the carbon skeleton and associated functional groups. The skeletal formulae for butane and methylpropane are shown in Figure 3 with other examples.

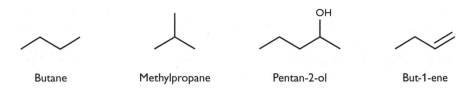

Butane   Methylpropane   Pentan-2-ol   But-1-ene

Figure 3

Cyclic compounds such as cyclohexane and benzene are represented as shown in Figure 4.

Knowledge check 3

Draw the skeletal formula for each of pent-2-ene, butan-2-ol and 3-chlorobut-1-ene.

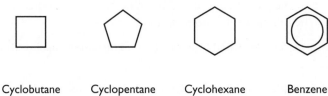

Cyclobutane   Cyclopentane   Cyclohexane   Benzene

Figure 4

Isomerism

Structural isomerism

Structural isomers are defined as compounds with the same molecular formula but different structural formulae. They occur with all functional groups, and you may be expected to both draw and name them. Figure 5 shows an example.

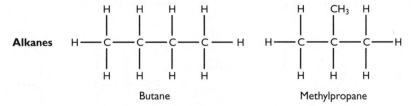

Alkanes

Butane   Methylpropane

Figure 5

E/Z (or cis–trans) isomerism

E/Z isomerism is found in alkenes. The key features to look for are:
- the C=C double bond
- each C in the C=C double bond is bonded to two different atoms or groups

The C=C double bond ensures that there is restricted rotation about a double bond, and the different atoms or groups attached to each carbon atom ensure that there is no symmetry around the carbon atom in the C=C double bond.

But-1-ene and but-2-ene both have a C=C double bond, but the right-hand carbon atom in the C=C double bond in but-1-ene is bonded to two hydrogen atoms and therefore does not exhibit E/Z (cis–trans) isomerism (Figure 6).

But-2-ene possesses both essential key features, and hence has a Z (cis) and an E (trans) isomer (Figure 7).

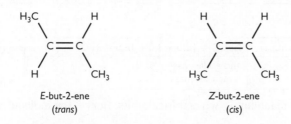

But-1-ene   But-2-ene

Figure 6

E-but-2-ene   Z-but-2-ene
(*trans*)   (*cis*)

E/Z isomers have a different geometry or 3D shape: this can be seen clearly when they are drawn as skeletal formulae

E-but-2-ene   Z-but-2-ene
(*trans*)   (*cis*)

Figure 7

> **Examiner tip**
> When explaining what is meant by *E/Z* isomerism most students remember that the C=C double bond restricts rotation but lots fail to mention the requirement that each C in the C=C double bond must be attached to two different atoms or groups.

Calculations

Most chemistry examinations will contain some calculations. The sort of calculations that could be tested within the organic chemistry section of Unit F322 are:

- empirical and molecular formulae calculations
- percentage yield calculations
- atom economy calculations

Empirical and molecular formulae calculations

The **empirical formula** is the simplest, whole-number ratio of atoms of each element in a compound.

The **molecular formula** is the actual number of atoms of each element in a molecule of a compound.

The question below is typical, and you would first be expected to calculate the empirical formula from the percentage composition by mass, and then to use the empirical formula and the relative molecular mass to deduce the molecular formula.

Example

Compound A has a relative molecular mass of 62.0 and has a composition by mass of C, 38.7%; H, 9.7%; O, 51.6%. Calculate the empirical formula and the molecular formula.

	C	:	H	:	O
Divide the % of each element by its own relative atomic mass	38.7/12.0	:	9.7/1.0	:	51.6/16.0
	3.2	:	9.7	:	3.2
Divide each by the smallest	3.2/3.2 = 1	:	9.7/3.2 = 3	:	3.2/3.2 = 1

Hence the empirical formula is $C_1H_3O_1 = CH_3O$.

Calculate the mass of the empirical formula	$CH_3O = 12.0 + 3.0 + 16.0 = 31.0$
Deduce how many empirical units are needed to make up the molecular mass	$\dfrac{\text{molecular mass}}{\text{empirical mass}} = \dfrac{62.0}{31.0} = 2$

Therefore, the molecular formula is made up of two empirical units. Hence, the molecular formula is $C_2H_6O_2$.

Percentage yield calculations

Percentage yield calculations involve mole calculations. If you are unsure of these, refer back to the Unit F321 guide in this series. A typical calculation is shown below.

Example

Ethanol, C_2H_5OH, can be oxidised to form ethanoic acid, CH_3COOH. If 2.3 g of ethanol are oxidised to produce 2.4 g of ethanoic acid, calculate the percentage yield.

Any mole calculations require a balanced equation, so it is essential that you are able to write suitable balanced equations and to use the mole ratios from the equation. [O] can be used to represent the oxidising agent.

Equation: $\quad C_2H_5OH + 2[O] \rightarrow CH_3COOH + H_2O$
Mole ratio: $\quad$ 1 mole 2 moles $\quad$ 1 mole $\quad$ 1 mole

The equation shows that 1 mole of ethanol produces 1 mole of ethanoic acid.

Step 1: calculate the number of moles of ethanol used:

amount of ethanol used =

$$n = \frac{\text{mass (in g)}}{\text{molar mass}} \qquad n = \frac{m}{M_r}$$

amount of ethanol used $= n = \dfrac{2.3}{46.0} = 0.05$ moles

Since the mole ratio is 1 : 1, the amount of ethanoic acid that could be made is also 0.05 moles.

Step 2: Calculate the number of moles of ethanoic acid actually produced:

$$\text{amount of ethanoic acid produced} = n = \frac{m}{M_r}$$

$$= \frac{2.4}{60.0} = 0.04 \text{ moles}$$

Step 3: calculate the percentage yield by using:

$$\text{percentage yield} = \frac{\text{actual yield} \times 100}{\text{maximum yield}}$$

$$= \frac{0.04 \times 100}{0.05}$$

$$= 80\%$$

Atom economy calculations

$$\text{atom economy} = \left(\frac{\text{molar mass of the desired products}}{\text{total molar mass of all the products}}\right) \times 100$$

The reaction between ethanol, C_2H_5OH, and ethanoic acid, CH_3COOH, is used to make the ester ethyl ethanoate, $CH_3COOC_2H_5$.

$$CH_3COOH + C_2H_5OH \rightarrow CH_3COOC_2H_5 + H_2O$$

The desired product is the ester, but water is also produced.

$$\text{molar mass of } CH_3COOC_2H_5 = 88.0\,g$$

$$\text{molar mass of } H_2O = 18.0\,g$$

$$\text{atom economy} = \left(\frac{88.0}{88.0 + 18.0}\right) \times 100$$

$$= \left(\frac{88.0}{106.0}\right) \times 100 = 83\%$$

A high atom economy is good, since that would indicate a low level of waste. Although concern over atom economy is a relatively new idea, its importance is likely to grow as society becomes ever more concerned about the need to conserve resources and reduce the production of unwanted by-products.

Reactions of functional groups

When describing the reactions of any functional group, you will be expected to:
- know the reagents
- know the conditions
- be able to write balanced equations
- be able to give the mechanism

Definitions

Reagents: these are chemicals involved in the reaction.

Conditions: these normally comprise the temperature, pressure and/or the use of a catalyst.

Knowledge check 4

A student reacted 5.00 g of ethanol (C_2H_5OH) with 8.00 g of ethanoic acid (CH_3COOH) and made 7.12 g of ethyl ethanoate ($CH_3COOC_2H_5$). Calculate the student's percentage yield.

Examiner tip

If asked to define atom economy, simply use this equation.

Examiner tip

When carrying out calculations do not round numbers during the calculation. Keep the number in your calculator and only round when you have finished the entire calculation.

Examiner tip

Words that end with '…phile' indicate a liking for. An 'anglophile' is someone who likes England. Don't be tempted to do this when explaining key terms like electrophile. You will *not* score any marks for writing that it 'loves electrons'. Stick to the scientific definitions given.

Mechanism: this breaks down the overall reaction into separate steps. It is usual to identify the attacking species. This is a radical, an electrophile or a nucleophile.

Radicals: reactive particles with an *unpaired* electron. The symbol for a radical generally shows the unpaired electron as a dot, for example $Cl\bullet$, $CH_3\bullet$.

Electrophiles: these are electron-deficient and can accept a pair of electrons — for example H^+, NO_2^+.

Nucleophiles: these are molecules or ions that can donate a lone pair of electrons — for example OH^-, NH_3.

Hydrocarbons: alkanes

Physical properties of alkanes

Hydrocarbons are compounds that contain hydrogen and carbon *only*. Alkanes and cycloalkanes are saturated hydrocarbons, as all the C–C bonds are single bonds. These bonds result in a tetrahedral shape, with bond angles of 109.5° around each carbon atom. Alkanes are obtained from crude oil, which is a complex mixture of hundreds of different hydrocarbons. The mixture undergoes initial separation by fractional distillation. Separation is achieved because the different alkanes have different boiling points.

The variation in boiling points depends on the amount of intermolecular bonding. Intermolecular bonding is covered in depth in the guide to Unit F321 of this series. There are essentially three possible types of intermolecular bonds:
- hydrogen bonds
- permanent dipole–dipole interactions
- van der Waals forces

The only intermolecular bonds found in alkanes are van der Waals forces.

If you are asked to explain or define a van der Waals force, there are three key features that you must include:
- the movement of electrons generates an instantaneous dipole
- the instantaneous dipole induces another dipole in neighbouring atoms or molecules
- the temporary attraction between the dipoles constitutes the van der Waals force

There are two important trends in the variation of boiling points in alkanes. First, as the relative molecular mass increases, the boiling point increases. This is due to:
- an increase in chain length
- an increase in the number of electrons

Both of the above result in an increase in the number of van der Waals forces.

Formula	Skeletal formula	Boiling point/°C
CH_4	n/a	−164
CH_3CH_3	—	−88
$CH_3CH_2CH_3$	∧	−42

Formula	Skeletal formula	Boiling point/°C
$CH_3CH_2CH_2CH_3$		0
$CH_3CH_2CH_2CH_2CH_3$		36
$CH_3CH_2CH_2CH_2CH_2CH_3$		69

Second, for isomers with the same relative molecular mass, the boiling points decrease with an increase in the amount of branching. This can be explained by the fact that straight chains pack closer together, creating more intermolecular forces. There are three isomers with the formula C_5H_{12} (Figure 8).

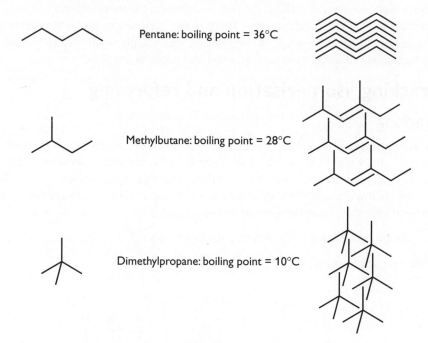

Pentane: boiling point = 36°C

Methylbutane: boiling point = 28°C

Dimethylpropane: boiling point = 10°C

Figure 8

Knowledge check 5
Put the following alkanes in order of boiling points starting with lowest boiling point first: octane, 2-methylpentane, hexane and 2,3-dimethylbutane.

Hydrocarbons: fuels

Crude oil as a source of organic chemicals

Fossilised deposits of natural gas and crude oil are the main source of alkanes. Natural gas and crude oil are known as **fossil fuels**.

Natural gas varies from one gas field to another, but is essentially methane mixed with other hydrocarbons such as ethane, propane and butane. It also usually contains hydrogen sulfide. Natural gas is processed and methane is separated from it before it is supplied to homes.

A **fossil fuel** is a non-renewable energy source. It is the product of the sun's energy, stored in plants millions of years ago through photosynthesis.

Crude oil

Crude oil is also a complex mixture of hydrocarbons and, like natural gas, the composition varies from place to place. Crude oil must be refined so that use can be made of the variety of hydrocarbons in the mixture. The initial separation is carried out by fractional distillation, which is a method for separating a mixture of liquids with different boiling points. The fractional distillation of crude oil is a continuous process and separates the crude oil into various fractions that contain a mixture of hydrocarbons within a narrow range of boiling points.

These fractions can be used directly or refined further and used as fuels or as feedstock for the petrochemical industry. The gasoline fraction (i.e. alkanes that contain about five to ten carbon atoms) is used as petrol. The naphtha fraction (i.e. alkanes that contain about six to ten carbon atoms) is the most important source of chemicals for the chemical industry. There is insufficient gasoline and naphtha to meet demand. Therefore, some of the heavier, less volatile fractions are treated to generate more of the high-demand gasoline and naphtha fractions.

Cracking, isomerisation and reforming

Cracking

Cracking involves heating the heavier oil fractions in the presence of a catalyst, so that long-chain molecules are broken down into shorter and more useful chains. The catalyst used can be a mixture of silicon and aluminium oxides or a zeolite. The process involves the random breaking of C–C and C–H bonds, and can lead to a variety of products. It can be summarised as:

long-chain alkane → shorter-chain alkane + alkene

Figure 9 shows an example.

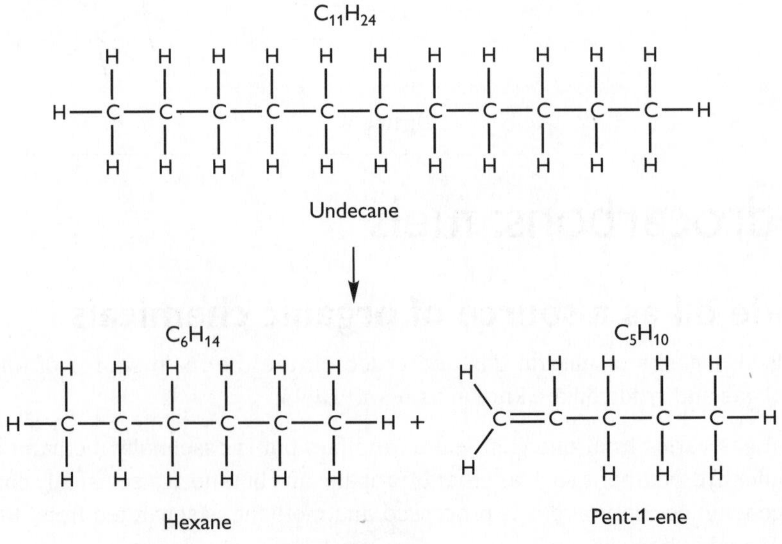

$C_{11}H_{24}$

Undecane

C_6H_{14}

Hexane

C_5H_{10}

Pent-1-ene

Figure 9

The bond breaking in the long-chain alkane is random, which results in a variety of products. The reaction scheme in Figure 10 shows how undecane, $C_{11}H_{24}$, can be cracked to produce a variety of different products.

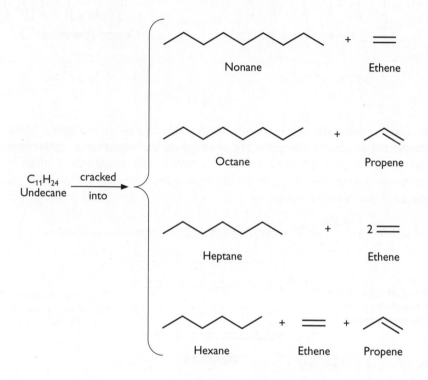

Figure 10

The alkenes produced in the cracking process are used as feedstock in the chemical industry to produce a wide variety of products. Alkenes are used as:
- monomers to make addition polymers; for example, ethene can be polymerised to poly(ethene), and propene can be used to produce poly(propene)
- feedstock in the production of alcohols; ethene can be converted into ethanol and propene into either propan-1-ol or propan-2-ol

Modern car engines require higher proportions of branched-chain alkanes, cycloalkanes and/or arenes. Straight-chain alkanes do not burn as smoothly or as efficiently as branched-chain alkanes, cycloalkanes or arenes. Straight-chain alkanes often pre-ignite, leading to 'knocking' or 'pinking', so they are processed into branched-chain alkanes, cycloalkanes or arenes.

Isomerisation

Isomerisation involves passing the vapour of straight-chain alkanes over a hot platinum catalyst (Figure 11).

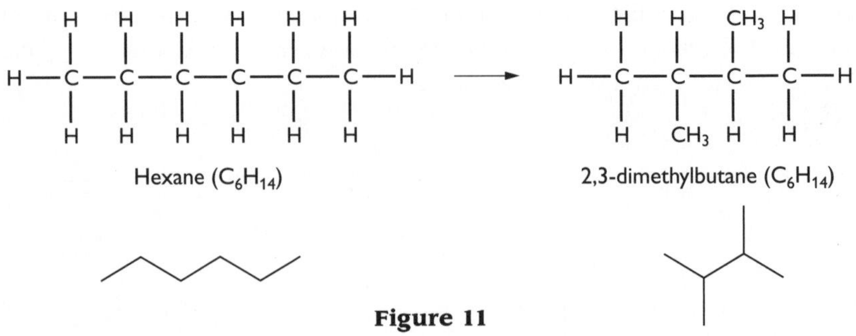

Hexane (C_6H_{14}) → 2,3-dimethylbutane (C_6H_{14})

Figure 11

The resulting mixture contains straight-chain and branched-chain alkanes. These are separated using a molecular sieve. The sieve is a zeolite, which has a regular network of pores (or holes). The size of the holes is critical. The straight-chain alkanes are able to pass through the zeolite, but the branched-chain isomers are too bulky; hence separation is achieved (Figure 12).

Zeolite

Branched-chain alkanes are too bulky to pass through the zeolite

Straight-chain alkanes can pass through the pores (holes) in the zeolite

Figure 12

Reforming

Reforming is used to obtain cycloalkanes and arenes from straight-chain alkanes. Reforming reactions are catalysed by bimetallic catalysts. The bimetallic catalyst usually consists of two transition elements, such as platinum and rhenium or platinum and iridium. Reforming always produces hydrogen, H_2, as a co-product.

The reforming of hexane into cyclohexane is shown in Figure 13.

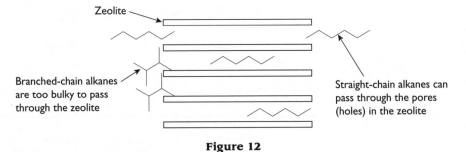

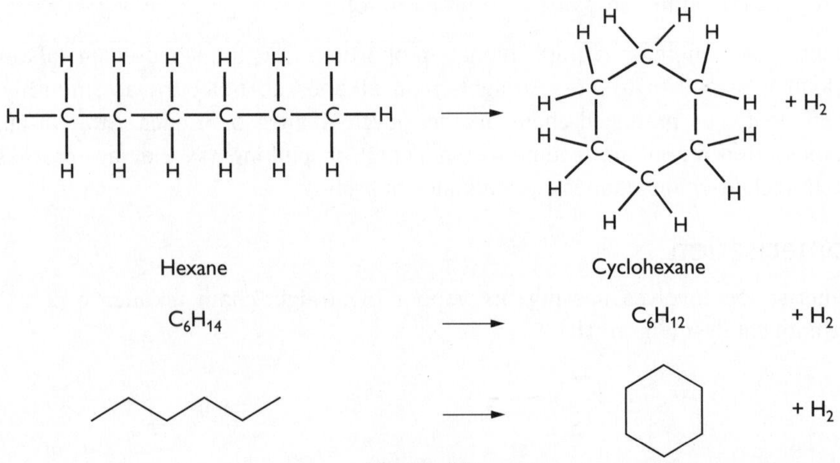

Hexane → Cyclohexane + H_2

C_6H_{14} → C_6H_{12} + H_2

+ H_2

Figure 13

The reforming of hexane, C_6H_{14}, can pass through four steps, resulting in the formation of benzene, C_6H_6. This is represented in skeletal form in Figure 14.

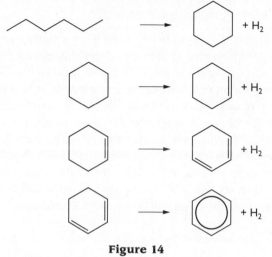

Figure 14

The overall equation is represented in Figure 15.

Figure 15

The three processes are summarised below.

Cracking

Long-chain alkanes are broken down into more useful, shorter-chain alkanes and alkenes. The alkenes are used to produce a wide range of useful products, such as alcohols and polymers.

Isomerisation

Straight-chain alkanes are turned into branched-chain alkanes, which burn more smoothly and are better and more efficient fuels.

Reforming

Straight-chain alkanes are turned into cycloalkanes or arenes, which burn more smoothly and are better and more efficient fuels. Hydrogen is always produced as a co-product. Branched-chain compounds, cycloalkanes and arenes are added to petrol to increase the octane rating.

Fuels and the future

Fossil fuels, such as gas and oil, are essential to society and are used extensively in industry, the home and in transport. However, our reserves of fossil fuels are finite

Knowledge check 6

Dodecane, $C_{12}H_{26}$, can be cracked in a variety of ways. Write balanced equations for the cracking of $C_{12}H_{26}$ into nonane and propene only and for the cracking of $C_{12}H_{26}$ into heptane, propene and ethene only.

and will eventually run out because they are non-renewable. When fossil fuels are exhausted, alternative sources of energy will be needed by society. In addition, the combustion of fossil fuels generates carbon dioxide, which is a greenhouse gas and is thought to be a key contributor to global warming.

Chemists and other scientists have been working for a number of years to develop alternative fuels. Currently, research is being carried out into ways of harnessing the energy in natural sources such as solar, tidal and wind energy. Additionally, time and effort are being put into the development of biofuels as well as nuclear fission and nuclear fusion. Biofuels such as bioethanol and biodiesel are made from crops such as wheat, sugar beet and vegetable oils. Both bioethanol and biodiesel are more environmentally friendly fuels, as they are classed as 'carbon-neutral'. Burning biofuels simply sends back into the atmosphere the carbon dioxide that the plants took out when they were growing in the fields. It sounds simple, but if biofuels are to replace fossil fuels then not only will car engines have to be adapted, but the need for more arable land to grow the crops will also have to be addressed.

Bioethanol

Bioethanol is made using yeast fermentation and can be mixed in small amounts with petrol and used in cars running on ordinary unleaded petrol. In the UK, bioethanol is already mixed with petrol and sold at a number of filling stations.

Biobutanol

Biobutanol is another alcohol-based fuel that can be mixed with petrol. Up to 10% biobutanol can be included in ordinary petrol and used by any car in the UK that runs on unleaded petrol. Biobutanol is made by fermentation, using bacterial enzymes.

Biodiesel

Biodiesel is produced from vegetable oil crops such as rape seed, or from the conversion of waste materials such as cooking oil. It can be mixed with conventional diesel.

Chemical reactions of alkanes

Alkanes are relatively unreactive, because:
- they are non-polar
- the C–C and the C–H bonds in alkanes are strong bonds

Combustion of alkanes

Alkanes burn easily and release energy (i.e. they undergo an exothermic reaction), and are used as fuels in industry, the home and in transport.

Complete combustion of alkanes in an excess of oxygen produces carbon dioxide and water.

$$CH_4 + 2O_2 \rightarrow CO_2 + 2H_2O$$

$$C_2H_6 + 3\tfrac{1}{2}O_2 \rightarrow 2CO_2 + 3H_2O$$

Incomplete combustion of alkanes in a limited supply of oxygen produces carbon monoxide and water.

$$CH_4 + 1\frac{1}{2}O_2 \rightarrow CO + 2H_2O$$
$$C_2H_6 + 2\frac{1}{2}O_2 \rightarrow 2CO + 3H_2O$$

Carbon monoxide is poisonous, and it is essential that hydrocarbon fuels are burnt in a plentiful supply of oxygen. Cars are fitted with catalytic converters (see pp. 52–53) to ensure that the amount of carbon monoxide emitted is reduced.

Substitution reactions of alkanes

Substitution by bromine and by chlorine to form halogenoalkanes

When describing the reactions of any functional group, you are expected to:
- know the reagents
- know the conditions
- be able to write balanced equations
- be able to give the mechanism

Reaction between methane and bromine

Reagent: Br_2

Conditions: ultraviolet light

Equation: $CH_4 + Br_2 \rightarrow CH_3Br + HBr$

Mechanism: the overall reaction is a **radical substitution**

Initiation: $Br_2 \rightarrow 2Br\bullet$

Propagation 1: $CH_4 + Br\bullet \rightarrow HBr + CH_3\bullet$

Propagation 2: $CH_3\bullet + Br_2 \rightarrow CH_3Br + Br\bullet$

Termination: any two free radicals

$CH_3\bullet + CH_3\bullet \rightarrow C_2H_6$ *or*

$CH_3\bullet + Br\bullet \rightarrow CH_3Br$ *or*

$Br\bullet + Br\bullet \rightarrow Br_2$

There are three distinct stages to the mechanism:

(1) Initiation: radicals are generated. The ultraviolet light provides sufficient energy to break the Br–Br bond homolytically and generates radicals. **Homolysis** or **homolytic fission** is when a covalent bond breaks so that the atoms joined by the bond separate and each atom takes one of the shared pair of electrons (Figure 16).

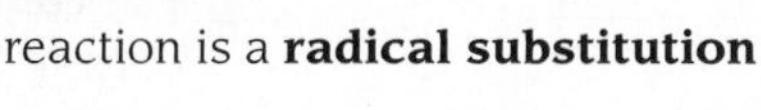

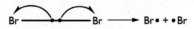

Figure 16

(2) Propagation: this involves two steps, each one maintaining the radical concentration. Usually, a chlorine or bromine radical is swapped for an alkyl radical or vice versa.

Examiner tip
Remember that the first propagation step *always* produces a hydrocarbon radical($\bullet C_nH_{2n+1}$)) and HBr (or HCl).

(3) Termination: this involves the loss of radicals.

Radical reactions have limitations, in that it is almost impossible to produce a single product. This is because radicals are very reactive and it is very difficult to avoid multiple substitutions of the hydrogen atoms in the alkane. Typically, for the reaction between CH_4 and Cl_2, the products would consist of CH_3Cl, CH_2Cl_2, $CHCl_3$ and CCl_4. This makes separation difficult and costly.

Hydrocarbons: alkenes

Alkenes and cycloalkenes are unsaturated hydrocarbons and contain at least one C=C double bond. The double bond is made up of a σ-bond and a π-bond.

A **σ-bond** is a single covalent bond made up of two shared electrons with the electron density concentrated between the two nuclei.

A **π-bond** is formed by the sideways overlap of two adjacent *p*-orbitals (Figure 17).

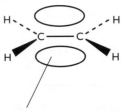

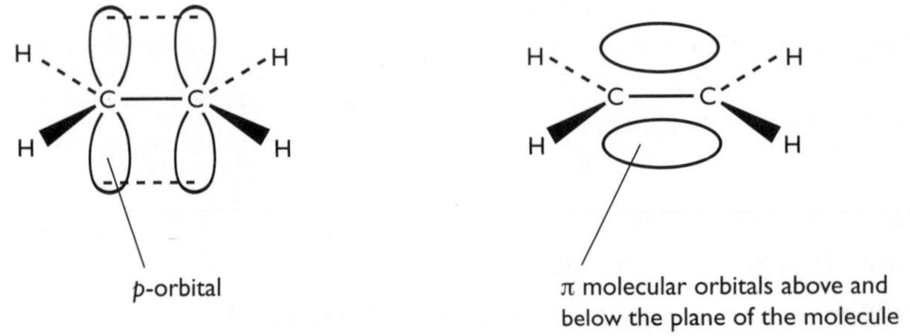

p-orbital

π molecular orbitals above and below the plane of the molecule

Figure 17

The bond angle at each side of the C=C double bond is approximately 120° (usually in the region 116–124°), which results in a trigonal planar structure around the C=C double bond (Figure 18).

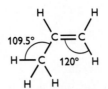

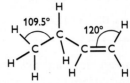

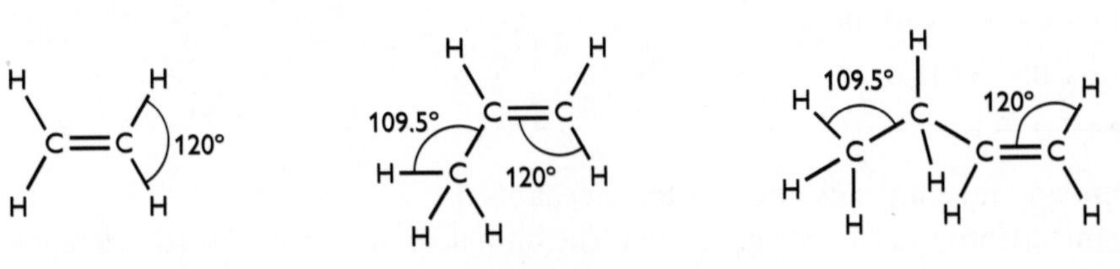

Figure 18

The C=C double bond prevents freedom of rotation, which under certain circumstances can lead to the existence of stereoisomers known as *E/Z (cis–trans)* isomers. The key features essential for *E/Z* isomers are:
- the C=C double bond
- each carbon atom in the C=C double bond is bonded to two different atoms or groups

Details of *E/Z* isomers are given on pp. 10–11.

It is important that you can recognise or predict whether or not *E/Z* isomers exist.

Examiner tip
If asked to describe, with the aid of a diagram, how the π-bond is formed in an alkene, students often draw the following which in fact shows a treble (not double) bond between the two carbons.

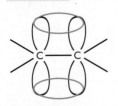

Examiner tip
When explaining *E/Z* isomerism makes sure you use the right wording. 'Each carbon atom in the C=C double bond is bonded to two different atoms or *molecules*' will not score the mark.

Reactions of alkenes

Alkenes are relatively reactive, because:
- the C=C bond consists of a σ-bond and a π-bond; the π-bond is weak and easily broken
- the C=C bond contains four electrons and has high electron density

The C=C double bond is an unsaturated bond and therefore undergoes addition reactions. Essentially, the double bond opens and an atom or group is added to each of the carbons. The general reaction can be summarised as:

$$CH_2=CH_2 + X–Y \rightarrow CH_2XCH_2Y$$

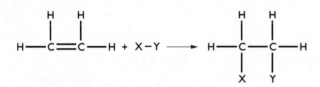

When preparing for the unit test, it is good practice to stick to a routine. For most organic reactions, it is useful to know:
- reagents
- conditions (if any)
- observations (if any)
- balanced equations

Hydrogenation
- Reagents: H_2
- Conditions: Ni catalyst, 150°C
- Observations: none
- Balanced equation: $CH_2=CH_2 + H_2 \rightarrow CH_3CH_3$

Hydrogenation of polyunsaturated vegetable oils derived from plants is used in the production of some margarines. Similar conditions are used, so that vegetable oils react with hydrogen in the presence of a nickel catalyst.

Bromination
- Reagents: Br_2
- Conditions: none
- Observations: decolorisation of bromine
- Balanced equation: $CH_2=CH_2 + Br_2 \rightarrow CH_2BrCH_2Br$

Formation of halogenoalkanes
- Reagents: HBr
- Conditions: none, but the HBr is made *in situ* from $NaBr + H_2SO_4$
- Observations: none
- Balanced equation: $CH_2=CH_2 + HBr \rightarrow CH_3CH_2Br$

Hydration
- Reagents: H_2O

Knowledge check 7
Write a balanced equation for the complete combustion of propene.

Examiner tip
When asked to describe what you would see when bromine reacts with an alkene, many students lose the mark by stating that the bromine would go 'clear'. Clear is the wrong word. Bromine is already clear — it is a clear red/brown liquid. When it reacts it loses its colour and the correct word to use is 'decolorised'.

Knowledge check 8
Write a balanced equation for the reaction between buta-1,3-diene and bromine, state what you would observe and name the organic product.

- Conditions: acid catalyst, such as H_3PO_4, 300°C, 6 MPa
- Observations: none
- Balanced equation: $CH_2=CH_2 + H_2O \rightarrow CH_3CH_2OH$

All other alkenes undergo similar reactions under similar conditions. However, unsymmetrical alkenes, like propene, can produce two isomers when reacted with hydrogen chloride (Figure 19) or water (Figure 20).

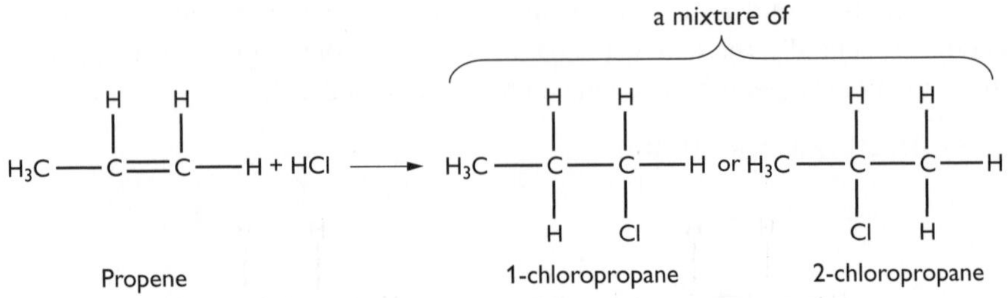

Figure 19

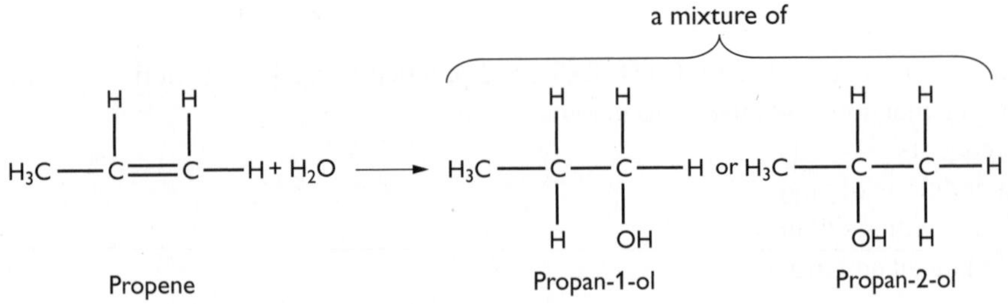

Knowledge check 9

Write a balanced equation for the hydration of but-2-ene.

Figure 20

Electrophilic addition

Mechanisms involving electrophiles and nucleophiles involve the movement of electron pairs. This movement is shown by the use of curly arrows. The curly arrow always points from areas that are electron-rich to areas that are electron-deficient.

When describing mechanisms, it is essential that you show:
- relevant dipoles
- lone pairs
- curly arrows

Alkenes, such as ethene, undergo electrophilic addition reactions. An electrophile is defined as a lone pair (of electrons) acceptor.

The key features to the mechanism are as follows:
- When the Br–Br approaches the ethene, a temporary induced dipole is formed, resulting in $Br^{\delta+}$–$Br^{\delta-}$.
- The initial curly arrow starts at the π-bond (within the C=C double bond) and points to the $Br^{\delta+}$.
- The second curly arrow shows the movement of the bonded pair of electrons in the Br–Br to the $Br^{\delta-}$, resulting in heterolytic fission of the Br–Br bond.
- The formation of an intermediate carbonium ion (also called a carbocation) and a :Br⁻ ion (that now contains the pair of electrons that were in the Br–Br bond) occurs.

- The third curly arrow from the :Br⁻ to the positively charged carbonium ion results in the formation of 1,2-dibromoethane.

The mechanism is best described using curly arrows, dipoles and relevant lone pairs of electrons (Figure 21).

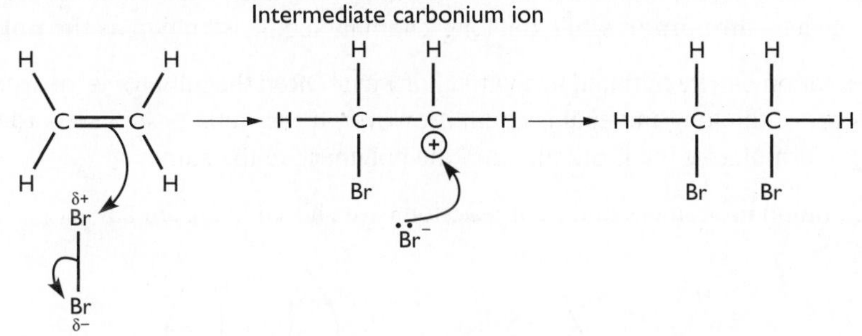

The intermediate ion can also be drawn as

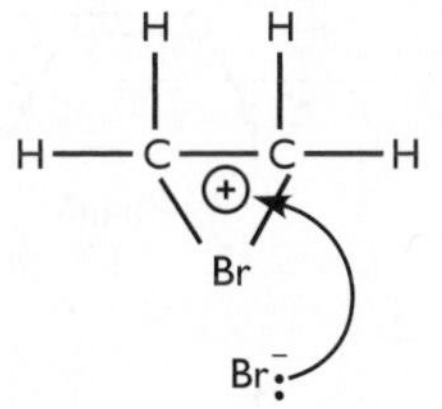

Figure 21

When Br_2 reacts with an alkene, the Br–Br undergoes heterolytic fission (Figure 22). Compare this to the way the Br–Br bond is broken when it reacts with an alkane (see p. 21).

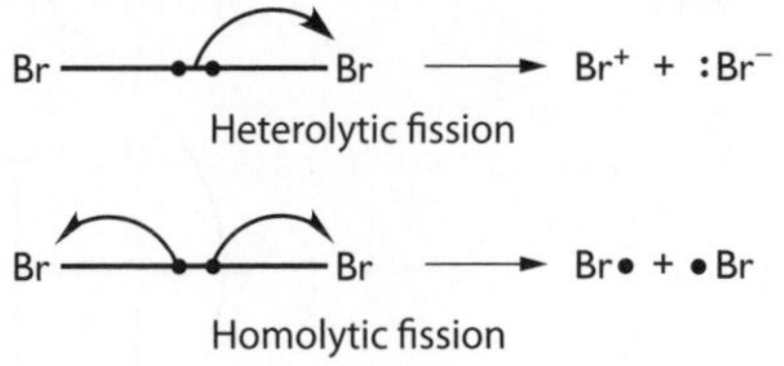

Figure 22

Examiner tip

Mechanisms are generally well understood but it is easy to lose lots of marks by rushing and/or being careless. Look at the response below for the mechanism of the reaction between propene and Br_2. At first glance it looks good *but* there are *seven* errors or omissions. Can you spot all seven?

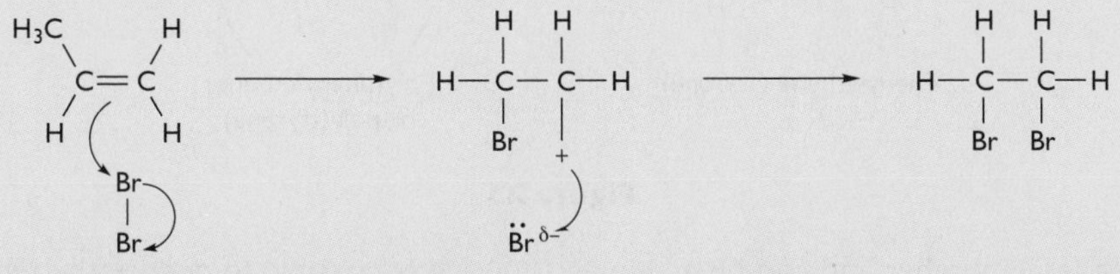

If you spot the errors made by others, it should prevent you from making the same mistakes.

Knowledge check 10

Explain what is meant by an *electrophile*. Write a balanced equation for the reaction between Br_2 and cyclohexene.

Polymerisation

Addition polymerisation

Alkenes can undergo addition reactions in which one alkene molecule joins to another and so on, until a long molecular chain is built up. The individual alkene molecule is referred to as a **monomer**, while the long-chain molecule is known as the **polymer**.

Polymerisation can be initiated in a variety of ways. Often the initiator is incorporated at the start of the long molecular chain. However, if the initiator is disregarded, the empirical formulae of the monomer and the polymer are the same.

Some common monomers and their reactions are shown in Figure 23.

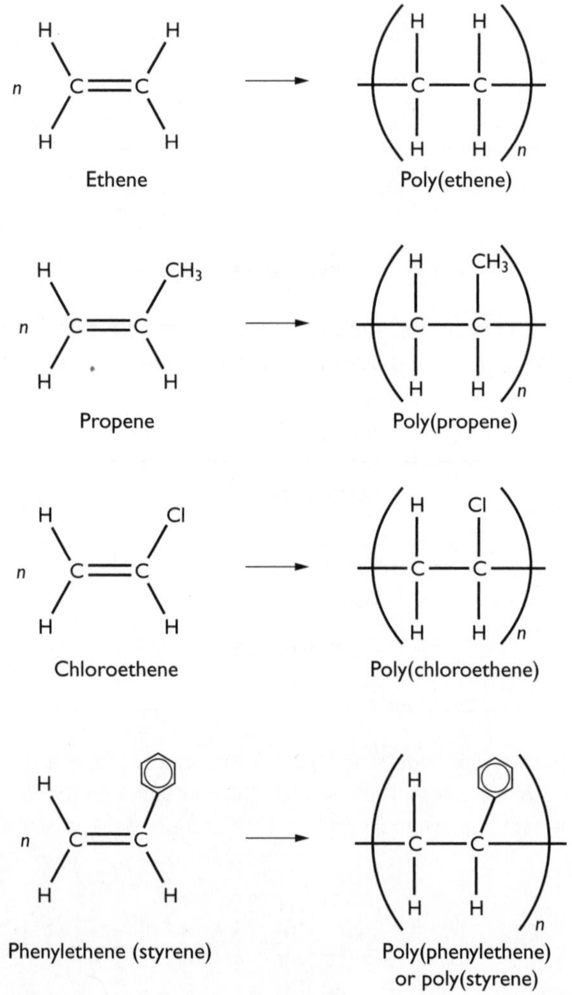

Ethene — Poly(ethene)

Propene — Poly(propene)

Chloroethene — Poly(chloroethene)

Phenylethene (styrene) — Poly(phenylethene) or poly(styrene)

Figure 23

n is a large number and can be as big as 10 000. It is possible to deduce the repeat unit of an addition polymer and to identify the monomer from which the polymer was produced (Figure 24).

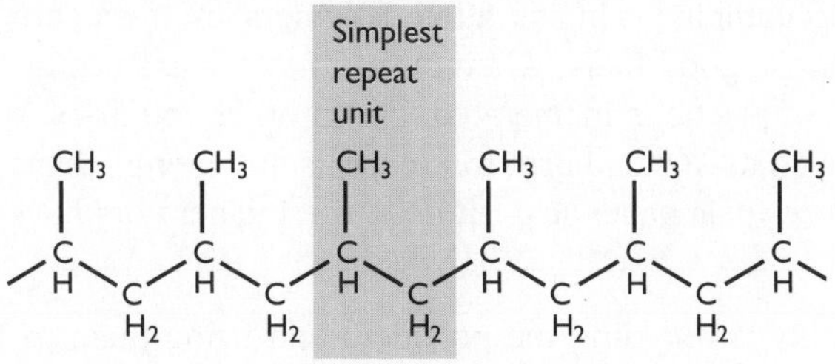

Simplest
repeat
unit

Therefore, the monomer is:

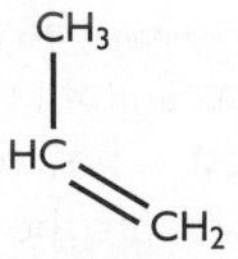

Propene

Figure 24

If asked to draw two repeat units of the polymer formed from propene, a common incorrect response is:

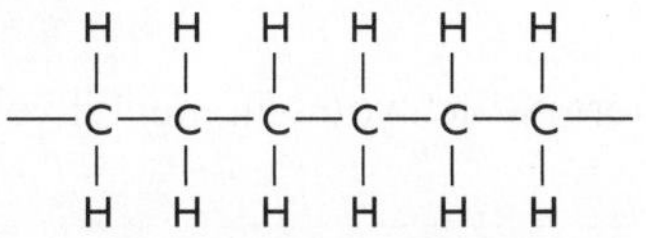

when it should be

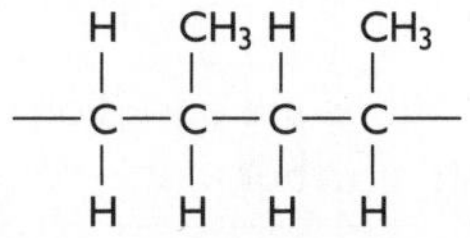

It is worth remembering that two repeat units of any addition polymer will always have a central backbone containing *four* carbon atoms.

The industrial importance of polymers

Polymers are an essential part of everyday life and have a wide variety of uses.

Polymer	Use
Poly(ethene)	Bags, insulation for wires, bottles
Poly(propene)	Plastic food boxes, clothing, ropes, carpets
Poly(chloroethene)	Clothes, drainpipes and guttering, window frames
Poly(styrene)	Packaging, telephones, flowerpots

Disposal of polymers

The widespread use of these polymers has created a major disposal problem. The bonds in addition polymers are strong covalent bonds and are non-polar, making most of the polymers resistant to chemical attack. As they are not broken down by bacteria, they are often referred to as being **non-biodegradable**.

Plastic waste has for many years been buried in landfill sites, where it remains unchanged for decades. This means that local authorities have to find more and more landfill sites.

Knowledge check 11

Write an equation to show the polymerisation of but-1-ene. Draw two repeat units of the polymer.

An alternative to dumping is incineration. Polymers are hydrocarbon-based and are therefore potentially good fuels. When burnt, they release useful energy. In the UK only about 10% of plastic is incinerated, but in other countries, such as Japan and Denmark, as much as 70% is burnt to create useful energy. Some plastics, such as PVC, also produce toxic gases (e.g. HCl), so the incinerators have to be fitted with gas-scrubbers.

Another possibility is recycling the polymers and using them as feedstock for the production of new polymers. Different types of polymer have to be separated from each other, as a mixture of polymers, when recycled, produces an inferior plastic product.

Biodegradable polymers are another possibility, and considerable effort has gone into trying to develop those that have suitable properties. The general principle is to create a polymer containing an active functional group that can be attacked by bacteria. Promising options are based on the polymerisation of isoprene, which is the monomer from which rubber is made. The systematic name for isoprene is 2-methylbuta-1,3-diene (Figure 25).

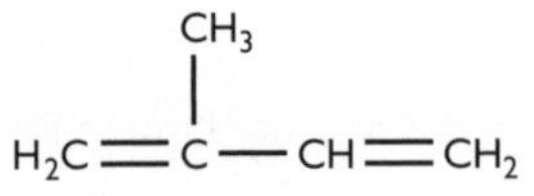

$$H_2C = \underset{\underset{CH_3}{|}}{C} - CH = CH_2$$

Figure 25

Other options are based on condensation polymers, which you will meet if you study chemistry further.

Summary

Having revised **Module 1: Basic concepts and hydrocarbons** you should now have an understanding of:

- the various ways in which organic formulae can be represented
- isomerism
- calculations used in organic chemistry
- key terms used in organic chemistry
- bonding, shape and boiling points of alkanes
- hydrocarbons as fuels including fractional distillation, cracking, isomerisation and reforming

- reactions of alkanes
- radical substitution mechanism
- bonding and shape of alkenes
- isomerism including *E*/*Z* isomers
- addition reactions
- electrophilic addition mechanism
- polymerisation and the environmental aspects of waste polymers

Module 2: Alcohols, halogenoalkanes and analysis

Alcohols

Properties

Alcohols all contain the hydroxyl group, –OH, and their names all end in '-ol'. They are classified as primary, secondary or tertiary alcohols.

Using R to represent any other attachment, we can identify the nature of the alcohol (Figure 26).

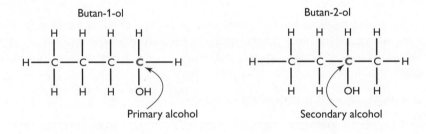

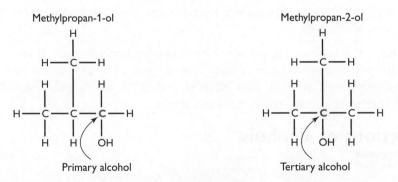

Figure 26

Alcohols have relatively high boiling points and are miscible with water. This can be explained in terms of hydrogen bonding.

Hydrogen bonds can be formed between the oxygen in the OH group in one alcohol and the hydrogen in the OH group in an adjacent alcohol (Figure 27).

Hydrogen bonding decreases the volatility, resulting in an increase in boiling point. Methanol and ethanol are freely miscible with water. When mixed, some of the hydrogen bonds in the separate liquids are broken, but they are then replaced by new hydrogen bonds between the alcohol and water. The higher the relative molecular mass of the alcohol, the lower is its miscibility with water.

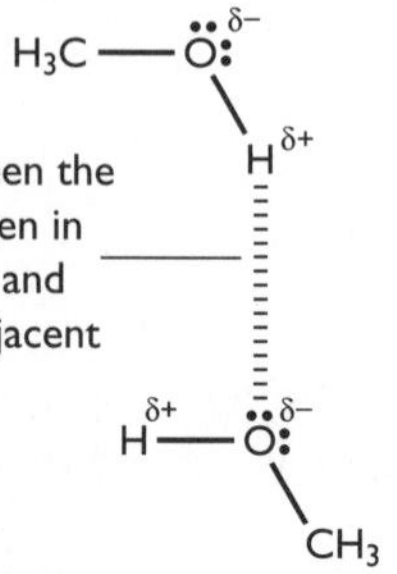

A hydrogen bond is formed between the lone pair of electrons on the oxygen in the O–H of one alcohol molecule and the hydrogen in the O–H of an adjacent alcohol molecule

Figure 27

The preparation of ethanol

Ethanol is widely used as a solvent. Most industrial ethanol is made by the addition reaction of steam with ethene in the presence of a phosphoric acid (H_3PO_4) catalyst.

$$CH_2{=}CH_2(g) + H_2O(g) \rightarrow CH_3{-}CH_2OH(g)$$

Ethene can be made as a co-product in the cracking of long-chain alkanes (see pp. 16–17).

Ethanol can also be produced by anaerobic fermentation of sugar found in fruit and grain. Fermentation involves yeasts, which occur naturally. The process is carried out in the absence of oxygen. Glucose (a sugar) is converted to ethanol and carbon dioxide.

$$C_6H_{12}O_6(aq) \rightarrow 2CH_3CH_2OH(aq) + 2CO_2(g)$$

Ethanol has been known to humans for thousands of years and is present in alcoholic drinks. Industrial alcohol, in the form of methylated spirits, is also used as a solvent. Increasingly, ethanol is being used as a biofuel and as a petrol substitute in countries with limited oil reserves.

Methanol is also used as a petrol additive to improve combustion and is important as a feedstock in the production of many organic chemicals, such as methanal (an essential component in MDF), ethanoic acid, and MTBE (an oxygenating additive used to improve the octane rating of the petrol mixture).

Reactions of alcohols

Examiner tip

When drawing alcohols you have to be careful how you draw the bond to the hydroxyl (OH) group. It is easy to lose the mark as shown in A and in B below. C shows how it should be drawn.

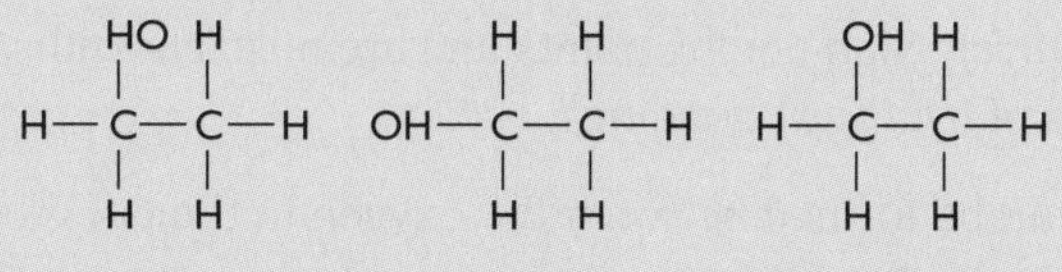

Examiner tip

Ethanol produced by fermentation is often incorrectly described as 're-usable' when the correct term should be 'renewable'. Renewable implies that when the ethanol is burnt it can be easily replaced because the feedstock, $C_6H_{12}O_6$, is obtained from plants. Clearly when the ethanol is burnt it forms CO_2 and H_2O and it cannot be re-used.

Combustion to produce carbon dioxide and water

- Reagents: excess oxygen
- Conditions: none
- Balanced equations:

$$CH_3OH + 1\tfrac{1}{2}O_2 \rightarrow CO_2 + 2H_2O$$

$$C_2H_5OH + 3O_2 \rightarrow 2CO_2 + 3H_2O$$

Dehydration to form an alkene

- Reagents: concentrated sulfuric acid or pumice/Al_2O_3
- Conditions: high temperature
- Balanced equation (Figure 28)

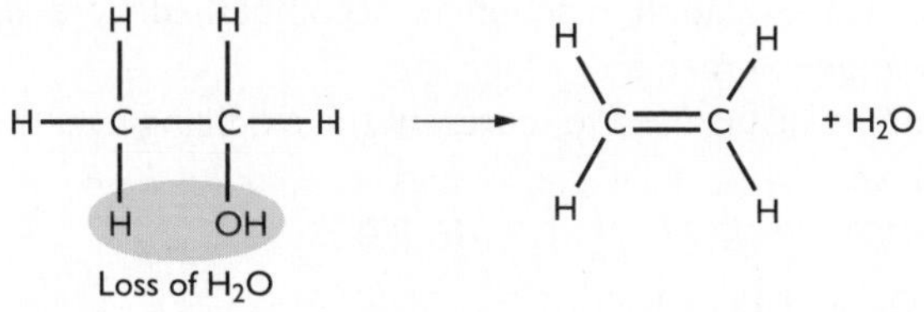

Figure 28

For alcohols like butan-2-ol it is possible to lose water in two ways (Figure 29).

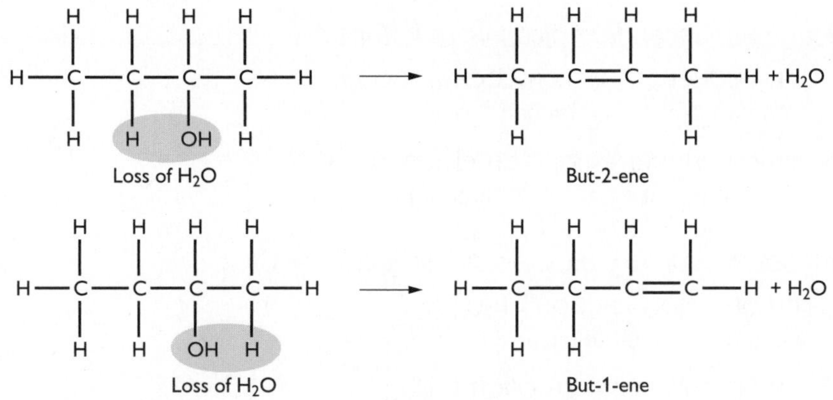

Figure 29

Esterification

Alcohols react with carboxylic acids in the presence of an acid catalyst to form esters.

- Reagents: carboxylic acid
- Conditions: acid catalyst
- Balanced equations:

$$CH_3OH + CH_3COOH \rightleftharpoons CH_3OOCCH_3 + H_2O$$

$$C_2H_5OH + CH_3COOH \rightleftharpoons C_2H_5OOCCH_3 + H_2O$$

Examiner tip

When writing equations for the combustion of alcohols many students forget to include the O in the alcohol and it is very common to see the equation for ethanol as:

$$C_2H_5OH + 3\tfrac{1}{2}O_2 \rightarrow 2CO_2 + 3H_2O$$

rather than:

$$C_2H_5OH + 3O_2 \rightarrow 2CO_2 + 3H_2O$$

Knowledge check 12

Write a balanced equation for the complete combustion of propan-1-ol.

Knowledge check 13

Write a balanced equation for the dehydration of pentan-3-ol.

Oxidation

Oxidation reactions differ, depending on whether the alcohol is primary, secondary or tertiary.

Combustion, dehydration and esterification reactions are common to alcohols, irrespective of whether they are primary, secondary or tertiary. Oxidation reactions differ depending on the classification of the alcohol.

- Primary alcohol → aldehyde → carboxylic acid
- Secondary alcohol → ketone
- Tertiary alcohol → resistant to oxidation
- Reagents: oxidising mixture $Cr_2O_7^{2-}/H^+$ (e.g. $K_2Cr_2O_7/H_2SO_4$)
- Conditions: when oxidising a primary alcohol, the choice of apparatus is important — refluxing will produce a carboxylic acid, while distillation produces an aldehyde
- Observations: each oxidation reaction is accompanied by a distinctive colour change from orange to green
- Balanced equations ([O] is used to represent the oxidising agent), see below.

Oxidation of primary alcohols to aldehydes

$$CH_3OH + [O] \rightarrow HCHO + H_2O$$
methanol   methanal

$$CH_3CH_2OH + [O] \rightarrow CH_3CHO + H_2O$$
ethanol   ethanal

Oxidation of secondary alcohols to ketones

$$CH_3CHOHCH_3 + [O] \rightarrow CH_3COCH_3 + H_2O$$
propan-2-ol   propan-2-one

$$CH_3CH_2CHOHCH_3 + [O] \rightarrow CH_3CH_2COCH_3 + H_2O$$
butan-2-ol   butanone

Oxidation of primary alcohols to carboxylic acids

$$CH_3OH + 2[O] \rightarrow HCOOH + H_2O$$
methanol methanoic acid

$$CH_3CH_2OH + 2[O] \rightarrow CH_3COOH + H_2O$$
ethanol   ethanoic acid

Halogenoalkanes

Halogenoalkanes are compounds in which one or more of the hydrogen atoms of an alkane have been replaced by a halogen atom. If one hydrogen has been replaced, the general formula is $C_nH_{2n+1}X$ (where X = F, Cl, Br or I).

Like alcohols, halogenoalkanes can be subdivided into primary, secondary and tertiary. The rules for classification are the same: if the carbon bonded to the halogen (X) is bonded to no more than one other carbon, then it is a primary halogenoalkane. However, if the carbon in the C–X is bonded to two other carbons, then it is a

---

Knowledge check 14

Write a balanced equation for the oxidation of butan-2-ol, use [O] to represent the oxidising agent.

Knowledge check 15

Explain what is meant by the terms *reflux* and *distillation*.

Examiner tip

Oxidation of primary alcohols to form carboxylic acids often leads to incorrect equations with the most common wrong response being:

$$CH_3CH_2OH + [O] \rightarrow$$
$$CH_3COOH + H_2$$

Remember that oxidation reactions *always* produce water as a product. The correct equation is:

$$CH_3CH_2OH + 2[O] \rightarrow$$
$$CH_3COOH + H_2O$$

OCR(A) AS Chemistry

secondary halogenoalkane. In a tertiary halogenoalkane, the carbon in the C–X is bonded to three other carbons.

The carbon–halogen bond is a polar bond and results in the carbon being susceptible to attack by a nucleophile. A nucleophile is defined as a lone pair (of electrons) donor.

Nucleophilic substitution

As with any mechanism, when describing nucleophilic substitution reactions, you should include:
- dipoles
- lone pairs of electrons
- curly arrows

Hydrolysis

- Reagents: NaOH or KOH (an alkali is required)
- Conditions: the solvent used must be water and the reaction mixture must be heated under reflux
- Balanced equation:

$$CH_3CH_2Br + NaOH \rightarrow CH_3CH_2OH + NaBr$$

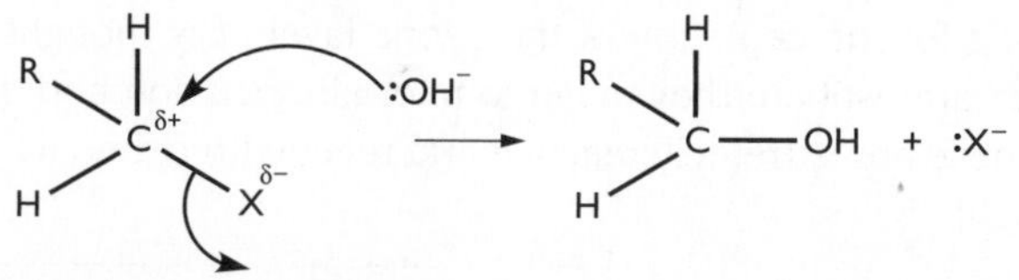

(R represents an alkyl group such as $CH_3–$, $C_2H_5–$ etc.)

Figure 30

The hydrolysis reaction (Figure 30) can be monitored by adding an aqueous ethanolic silver nitrate solution. The substituted halide ion reacts with the Ag^+ ion, producing precipitates of one of the following:
- white AgCl(s)
- cream AgBr(s)
- yellow AgI(s)

Rate of hydrolysis

When equal amounts of 1-chlorobutane, 1-bromobutane and 1-iodobutane are reacted, under identical conditions, with a hot aqueous ethanolic solution of silver nitrate, it can be clearly seen that 1-iodobutane reacts the fastest and 1-chlorobutane the slowest. This can be explained by comparing the carbon–halogen bond enthalpies.

Bond	Bond enthalpy (kJ mol^{-1})
C–F	467
C–Cl	340
C–Br	280
C–I	240

Clearly, the C–I bond is the weakest and therefore requires least energy to break it. The C–F is so strong that it rarely, if ever, undergoes hydrolysis.

Uses of halogenoalkanes

Halogenoalkanes are extremely useful compounds. They are involved in the preparation of a wide range of products, including pharmaceuticals (such as ibuprofen); polymers such as PVC, which is made from $CH_2=CHCl$; and PTFE which is made from $F_2C=CF_2$.

CFCs

CFCs such as dichlorodifluoromethane, CCl_2F_2, and trichlorofluoromethane, CCl_3F, were developed to be used in air conditioning, refrigeration units and aerosols, as well as being used as blowing agents in the production of foamed polymers such as expanded polystyrene. They are suitable for these purposes because they are unreactive, non-flammable and non-toxic, as well as being liquids of low volatility that can be readily evaporated and re-condensed. However, it is also these properties that make them so persistent in the atmosphere. Our knowledge at the time of their introduction did not extend to understanding the dangerous effect they would ultimately have in the stratosphere.

Knowledge check 16

Why is the C–Cl bond broken in CCl_2F_2 rather than the C–F bond?

CFCs are blamed for the depletion of the ozone layer. It is thought that when CFCs reach the upper atmosphere they undergo photodissociation and generate chlorine radicals, Cl•. These are extremely reactive and react with ozone (O_3) in the presence of ultraviolet light.

$$CCl_2F_2 \rightarrow \bullet CClF_2 + Cl\bullet$$

The chlorine radical is then involved in the propagation steps:

$$Cl\bullet + O_3 \rightleftharpoons ClO\bullet + O_2$$

$$ClO\bullet + O \rightleftharpoons Cl\bullet + O_2$$

The Cl• is regenerated in the second propagation step and can then go on to react with other ozone molecules.

Chemists are working to minimise damage to the environment by researching alternatives to CFCs. Initially, these centred on the use of HCFCs such as 1,1,1,2-tetrafluoroethane, which include a C–H bond in their structure, making them more degradable in the atmosphere. Currently, hydrocarbons are used as alternative propellants in aerosols. Carbon dioxide has been found to be a suitable alternative blowing agent in the creation of expanded polystyrene.

Modern analytical techniques

Infrared spectroscopy

Infrared spectra can be used to identify key absorptions of the alcohol, carbonyl, carboxylic acid and amine functional groups. (Data sheets will be supplied in all examinations, so there is no need to learn these values.)

Bond	Location	Wavenumber/cm⁻¹
C–O	Alcohols, esters, carboxylic acids	1000–1300
C=O	Aldehydes, ketones, carboxylic acids, esters, amides	1640–1750
C–H	Any organic compound with a C–H bond	2850–3100
O–H	Carboxylic acids	2500–3300 (very broad)
O–H	Alcohols, phenols	3200–3550 (broad)
N–H	Amines, amides	3200–3500

Amines, amides and phenols will only appear in A2 chemistry, but you might be expected to distinguish between alcohols and their oxidation products: aldehydes, ketones and carboxylic acids.

For an alcohol you need to identify *two* peaks (Figure 31):
- 1000–1300 cm⁻¹: all alcohols contain a C–O bond
- 3200–3500 cm⁻¹: all alcohols contain an O–H bond, which is a broad peak and should not be confused with the small, sharp peaks due to C–H bonds

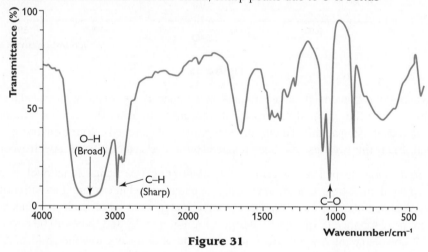

Figure 31

For an aldehyde or a ketone, you need to identify just *one* peak (Figure 32):
- 1640–1750 cm⁻¹: all aldehydes and ketones contain a C=O bond

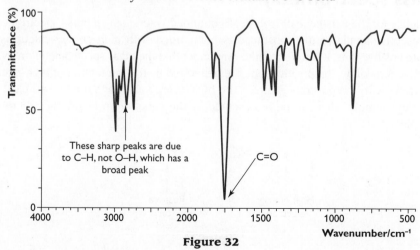

Figure 32

For a carboxylic acid you need to identify *two* (possibly three) peaks (Figure 33):

- $1640-1750 \, cm^{-1}$: all carboxylic acids contain a C=O bond
- $2500-3300 \, cm^{-1}$: all carboxylic acids contain an O–H bond, which is very broad
- $1000-1300 \, cm^{-1}$: all carboxylic acids contain a C–O bond

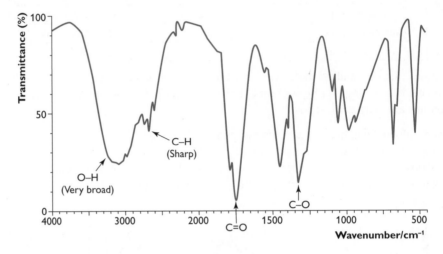

Figure 33

Nowadays, modern analytical techniques have replaced 'wet-tests' or chemical tests. In 1967, when the breathalyser was first introduced into the UK, alcohol (ethanol) was detected using acidified dichromate crystals that changed colour from orange to green. Today this 'wet-test' has been replaced by analysis using infrared spectroscopy.

Infrared spectroscopy is a powerful tool in identifying a particular functional group, but when identifying a specific chemical it is usually used in conjunction with other analytical techniques, such as mass spectrometry, chromatography and nuclear magnetic resonance (NMR) spectroscopy. Chromatography and NMR will be covered in A2 chemistry. At AS you are expected to be able to link together information obtained from an infrared spectrum and a mass spectrum.

Mass spectrometry

Evidence for the existence of isotopes is obtained when using a mass spectrometer. A gaseous sample of the substance is bombarded with high-energy electrons to create positive ions. It is important to remember that the mass spectrometer analyses positive ions only. The positive ions are generated by the interaction of high-energy electrons with the gaseous sample. The high-energy electrons remove an electron from the sample. This can be represented by the equation in Figure 34, where the sample is X.

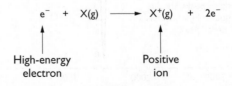

Figure 34

OCR(A) AS Chemistry

The positive ions are passed through a magnetic field, where they are deflected. The amount of deflection is dependent on their mass. A detector is calibrated to record the degree of deflection and to interpret this in terms of the mass of the ions.

For an atom, a typical printout from the detector looks like a 'stick-diagram', with each stick representing an ion (isotope). The bigger the 'stick', the more abundant is the ion.

The mass spectrum for boron is shown in Figure 35.

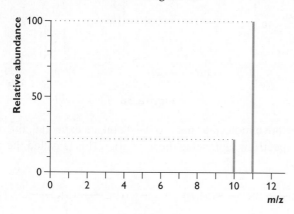

Figure 35

The x-axis is labelled 'm/z'. This means mass/charge, but since the charge is 1+ it is effectively the relative mass of the ions that is recorded. It is calibrated so that the isotope carbon-12 is given a value of exactly 12. This is the standard that is universally used for the comparison of the masses of atoms.

The mass spectrum of boron shows two lines, indicating that there are two isotopes with masses of 10 and 11. The relative abundance of these two isotopes is shown by the heights of the lines — they are in the ratio of 23 (boron-10) to 100 (boron-11).

$$\text{relative atomic mass} = \frac{(10 \times 23) + (11 \times 100)}{123} = 10.8$$

The mass spectrometer is able to work with very small samples of gaseous molecules, and can be operated remotely. This has made it an extremely useful piece of equipment for analysing samples in remote locations. The equipment on board the Mars space probe included a mass spectrometer that was able to analyse gaseous samples and return the results to Earth.

The mass spectrometer can also analyse compounds, although the number of lines obtained is usually quite large. This is because bombardment by electrons causes the molecule to break up, with each of the fragments obtained registering on the detector. This can be an advantage, as it is sometimes possible to obtain details of the structure of the molecule as well as its overall molecular mass.

Figure 36 is the mass spectrum of propane (C_3H_8).

Base peak
Molecular ion
Relative abundance
100
80
60
40
20
0
0
10
20
30
40
50
m/z

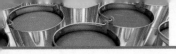

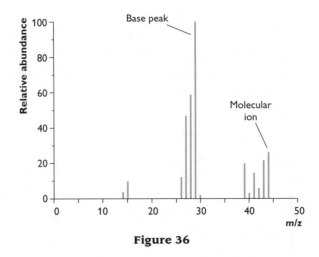

Figure 36

Propane has a relative molecular mass of 44.0 and, as expected, the peak furthest to the right of the spectrum represents the ion $C_3H_8^+$. This is called the **molecular ion peak** (Figure 37).

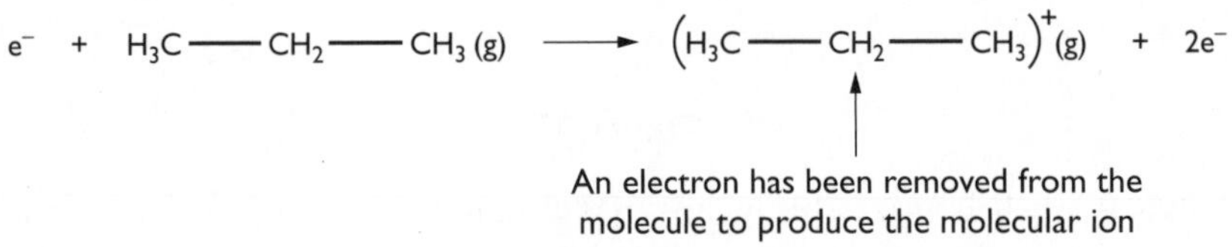

e⁻ + H₃C—CH₂—CH₃ (g) → (H₃C—CH₂—CH₃)⁺ (g) + 2e⁻
An electron has been removed from the molecule to produce the molecular ion

Figure 37

However, bombardment by electrons also creates ions of fragments of the molecule. These ions will also be registered on the printout. The peak at 29 occurs because a CH_3 unit has been broken from the $CH_3CH_2CH_3$ chain, and thus the ion $CH_3CH_2^+$ has been detected (Figure 38).

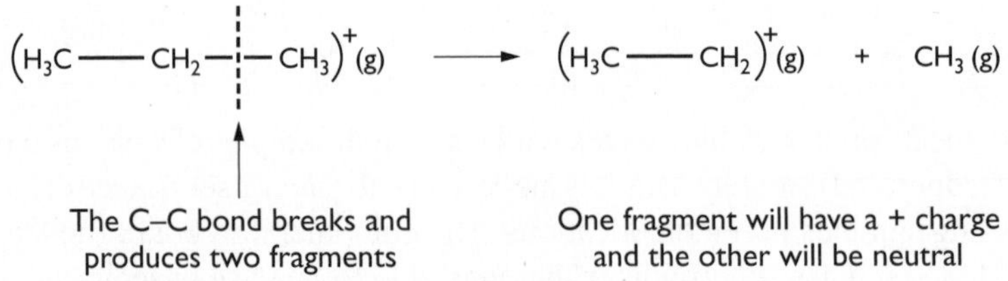

(H₃C—CH₂—CH₃)⁺ (g) → (H₃C—CH₂)⁺ (g) + CH₃ (g)
The C–C bond breaks and produces two fragments
One fragment will have a + charge and the other will be neutral

Figure 38

Similarly, the peak at 15 represents a CH_3^+ ion (Figure 39) and it is possible to suggest the identity of all other peaks in the spectrum.

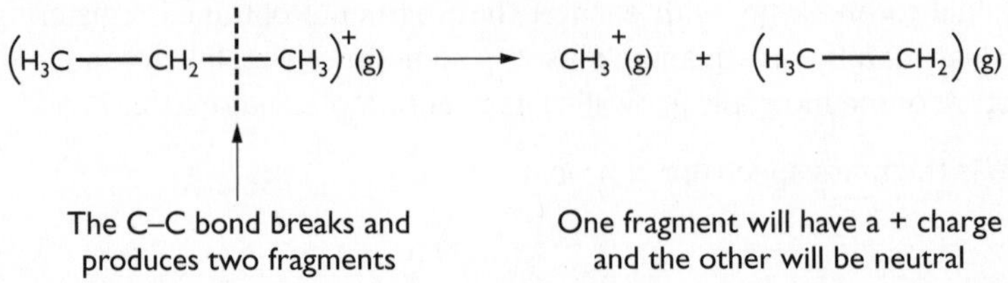

(H₃C—CH₂—CH₃)⁺ (g) → CH₃⁺ (g) + (H₃C—CH₂) (g)
The C–C bond breaks and produces two fragments
One fragment will have a + charge and the other will be neutral

Figure 39

Fragmentation leads to a large number of peaks, and the pattern of these is often said to be a fingerprint of the molecule. Like a fingerprint, it can be used in conjunction with a computer containing a spectral database to identify a particular chemical.

Butan-1-ol and butan-2-ol have very similar infrared spectra. Their mass spectra are shown in Figure 40.

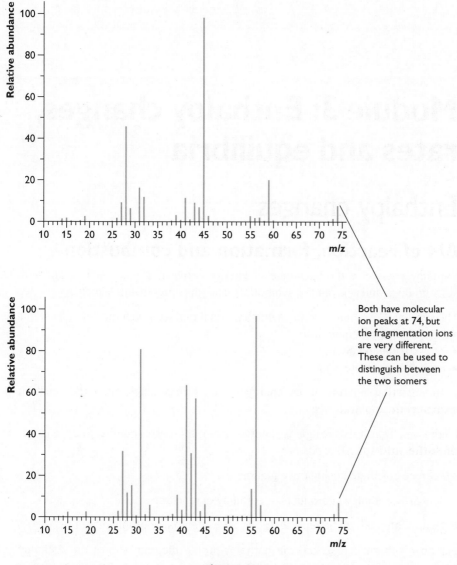

Both have molecular ion peaks at 74, but the fragmentation ions are very different. These can be used to distinguish between the two isomers

Figure 40

Analysis of the fragmentation of molecules can be used to distinguish between structural isomers. Although this may not always be easily apparent at a glance, computers can be used to compare the pattern obtained from a compound with a database of spectra, and this is a widely used research tool.

Knowledge check 17

In the mass spectrum of butan-2-ol suggest the identity of peaks with m/z values of: 15, 45, 59 and 74.

Summary

Having revised **Module 2: Alcohols, halogenoalkanes and analysis** you should now have an understanding of:

- industrial preparation of ethanol by either fermentation or by hydration
- reactions of alcohols including oxidation, elimination and esterification

- hydrolysis of halogenoalkanes
- nucleophilic substitution mechanisms
- CFCs and ozone
- how to recognise absorptions due to O–H and C=O bonds in infrared spectra
- how to determine the molar mass of molecules by using the molecular ion peak in mass spectra

Module 3: Enthalpy changes, rates and equilibria

Enthalpy changes

Δ*H* of reaction, formation and combustion

Enthalpy change is the exchange of energy between a reaction mixture and its surroundings and is given the symbol ΔH. The units are always $kJ\,mol^{-1}$.

The symbol 'Δ' is often used in chemistry, and it indicates a change in…, such that:
- ΔT = change in temperature
- ΔV = change in volume
- ΔP = change in pressure

If the reaction mixture loses energy to its surroundings, then the reaction is **exothermic** and ΔH is *negative*.

If the reaction mixture gains energy from its surroundings, the reaction is **endothermic** and ΔH is *positive*.

ΔH can be calculated using the equation:

ΔH = enthalpy of products — enthalpy of reactants

Enthalpy changes can be represented by simple enthalpy-profile diagrams.

For an exothermic reaction, the enthalpy-profile diagram shows the enthalpy of products at a lower energy than the enthalpy of reactants. For an endothermic reaction, the enthalpy-profile diagram shows the enthalpy of products at a higher energy than the enthalpy of reactants. The difference in the enthalpy is ΔH. E_a is the activation energy (Figure 41).

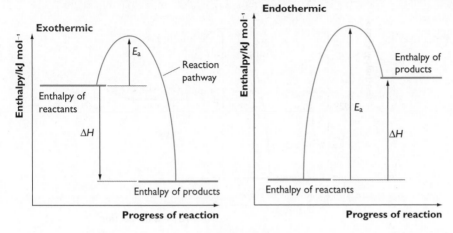

Figure 41

Oxidation reactions such as the combustion of fuels are exothermic and release energy to their surroundings. This results in an increase in temperature in the surroundings. The enthalpy profile in Figure 42 illustrates the combustion of methane.

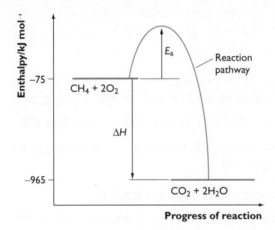

Figure 42

Oxidation of carbohydrates, such as glucose, in respiration is also an exothermic reaction.

Thermal decomposition reactions are usually endothermic, requiring energy from the surroundings. The enthalpy profile for the thermal decomposition of calcium carbonate is shown in Figure 43.

Enthalpy-profile diagrams not only show the enthalpy change of reaction, ΔH, but also display the activation energy, E_a.

Activation energy is defined as the minimum energy required, in a collision between particles, if they are to react. (This is covered in more detail in the section on reaction rates; see p. 48.) In any chemical reaction, bonds have to be broken and new bonds have to be formed. Breaking bonds is an endothermic process, requiring energy. This energy requirement contributes to the activation energy of a reaction.

Knowledge check 18

Define the term activation energy E_a.

Draw an energy-profile diagram to show how a catalyst works.

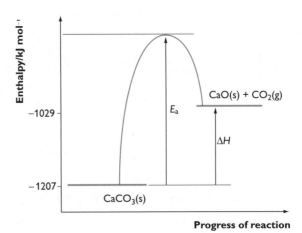

Figure 43

Standard enthalpy changes

All standard enthalpy changes are measured under standard conditions. The temperature and the pressure at which measurements and/or calculations are carried out are standardised.

Standard temperature = 298 K (25°C)

Standard pressure = 101 kPa (100 000 N m^{-2} = 10^5 Pa = 1 bar = 1 atmosphere)
(For examination purposes, 100 kPa is acceptable.)

Standard temperature and pressure are often referred to as STP.

Examinations often ask for a definition of enthalpy changes, and it is advisable to learn the following definitions.

Standard enthalpy change of reaction

ΔH_r° is the enthalpy change when the number of moles of the substances in the balanced equation react under the standard conditions of 298 K and 100 kPa.

Standard enthalpy change of formation

ΔH_f° is the enthalpy change when one mole of a substance is formed from its elements, in their natural state, under the standard conditions of 298 K and 100 kPa.

Standard enthalpy change of combustion

ΔH_c° is the enthalpy change when one mole of a substance is burnt completely, in an excess of oxygen, under the standard conditions of 298 K and 100 kPa.

Average bond enthalpy

This is the enthalpy change on breaking one mole of a covalent bond in a gaseous molecule under the standard conditions of 298 K and 100 kPa.

In addition to these definitions, you may also be expected to show your understanding by writing equations to illustrate the standard enthalpy change of reaction, formation and combustion.

The equation for the standard enthalpy of neutralisation of HCl(aq) with NaOH(aq) is:

$$HCl(aq) + NaOH(aq) \rightarrow NaCl(aq) + H_2O(l)$$

An ionic equation can also be used to show the standard enthalpy of neutralisation:

$$H^+(aq) + OH^-(aq) \rightarrow H_2O(l)$$

Equations to show the standard enthalpy of formation of a substance must reflect the standard definition and must always:
- show the elements as the reactants
- produce *one* mole of the substance, even if that means having fractions in the equation
- show the state symbols

The equation for the standard enthalpy of formation of:
- ethane is $2C(s) + 3H_2(g) \rightarrow C_2H_6(g)$
- ethanol is $2C(s) + 3H_2(g) + \frac{1}{2}O_2(g) \rightarrow CH_3CH_2OH(l)$

Equations to show the standard enthalpy of combustion of a substance must reflect the standard definition and must always:
- react *one* mole of the substance with excess $O_2(g)$, even if that means having fractions in the equation
- show the state symbols

Usually, the products are $CO_2(g)$ and $H_2O(l)$.

The equation for the standard enthalpy of combustion of:
- ethane is $C_2H_6(g) + 3\frac{1}{2}O_2 \rightarrow 2CO_2(g) + 3H_2O(l)$
- ethanol is $CH_3CH_2OH + 3O_2 \rightarrow 2CO_2(g) + 3H_2O(l)$

You will be expected to calculate enthalpy changes:
- using enthalpy data from experimental data
- average bond enthalpies
- by using Hess's law

Enthalpy changes using enthalpy data from experiments

The standard enthalpy change, ΔH_r°, for reactions that take place in solution can usually be measured directly by using the simple apparatus shown in Figure 44. However, the results obtained will only be approximate, since there are likely to be substantial heat losses to the surroundings.

Figure 44

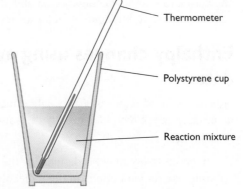

Thermometer

Polystyrene cup

Reaction mixture

Knowledge check 20

Define standard enthalpy change of formation and write an equation to illustrate the standard enthalpy change of formation of propanone, $CH_3COCH_3(l)$.

Knowledge check 21

Define standard enthalpy change of combustion and write an equation to illustrate the standard enthalpy change of combustion of propanal, $CH_3CH_2CHO(l)$.

The energy transfer for the reaction mixture is given the symbol Q and can be calculated by using the equation:

$$Q = mc\Delta T$$

where m is the mass of the reaction mixture, c is the specific heat capacity of the reaction mixture and ΔT is the change in temperature.

The solvent used for most reactions is water. The specific heat capacity of water, c, is taken as either $4.2\,J\,g^{-1}\,K^{-1}$ or $4.2\,kJ\,kg^{-1}\,K^{-1}$. The enthalpy change for the reaction mixture will give a value in either J (joules) or kJ (kilojoules), depending on the value/ units of the specific heat capacity. It is usual to correct this value (Q), so that the ΔH value can be quoted for one mole of reactant and the units of ΔH become $kJ\,mol^{-1}$. The standard enthalpy change for the reaction can then be calculated by dividing the energy transferred by the number of moles, n, of reactant used.

$$\Delta H_r = \frac{Q}{n} = \frac{mc\Delta T}{n}$$

Example

When $50.0\,cm^3$ of $2.00\,mol\,dm^{-3}$ hydrochloric acid was mixed with $50.0\,cm^3$ of $2.00\,mol\,dm^{-3}$ sodium hydroxide, the temperature increased by $13.7°C$. Calculate the standard enthalpy change of neutralisation of HCl(aq). Assume that the specific heat capacity, c, is $4.20\,J\,g^{-1}\,K^{-1}$ and that the densities of the HCl and NaOH are both $1.00\,g\,cm^{-3}$.

Method

Step 1: calculate the energy transferred in the reaction by using the equations

$$Q = mc\Delta T$$
m = the mass of the two solutions = 50 + 50 = 100 g
$Q = mc\Delta T = 100 \times 4.20 \times 13.7 = 5754\,J = 5.754\,kJ$

Step 2: convert the answer to $kJ\,mol^{-1}$ by dividing by the number of moles used

the amount in moles of HCl, $n = cV = 2.00 \times \dfrac{50}{1000} = 0.1\,mol$

$$\Delta H = \frac{Q}{n} = \frac{5.754}{0.1} = 57.54 = 57.5\,kJ\,mol^{-1}$$

Remember that, because the temperature increased, it is an exothermic reaction and therefore $\Delta H = -57.5\,kJ\,mol^{-1}$.

The standard enthalpy change of combustion, ΔH_c°, for a volatile liquid can also be measured directly, but again heat losses mean that the result is only approximate.

Enthalpy changes using average bond enthalpy data

Breaking a bond *requires energy*, while making a bond *releases energy*. The energy required to break a bond is exactly the same as the energy released when the same bond is made.

The bond enthalpy may be defined as the enthalpy change required to break and separate one mole of bonds in the molecules of a gas, so that the resulting gaseous

Examiner tip

In a question such as:

2.30 g ethanol was burnt and raised the temperature of 300 cm³ water by 12.5°C. Calculate ΔH_c for ethanol.

you would be told the density of the water as well as the specific heat capacity, c, of the apparatus. Many make an error in the first step when using $Q = mc\Delta T$ and incorrectly use 2.30 g as the mass. The mass should be 300 g, which is the mass of the water.

(neutral) particles, atoms or radicals exert no forces upon each other. It is best reinforced by a simple equation such as:

$$Cl–Cl(g) \rightarrow Cl\bullet(g) + Cl\bullet(g)$$

or generally:

$$X–Y(g) \rightarrow X\bullet(g) + Y\bullet(g)$$

Bond enthalpies are the average (mean) values, and do not take into account the specific chemical environment.

Some average bond enthalpies are shown in the table below.

Bond	C–H	C=O	O=O	O–H	C–N
ΔH/kJ mol^{-1}	+413	+805	+498	+464	+286
Bond	C=C	N≡N	N–H	C–C	H–H
ΔH/kJ mol^{-1}	+612	+945	+391	+347	+436

Calculations involving average bond enthalpy

These calculations are straightforward. In order for a reaction to take place, existing bonds have to be broken (this needs an input of energy, i.e. it is endothermic), and then new bonds have to be formed (energy is given out, so this is exothermic). For a simple reaction involving gaseous molecules:

$$\Delta H = \Sigma \text{ bond enthalpies of reactants} - \Sigma \text{ bond enthalpies of products}$$

The enthalpy of combustion of methane can be calculated using average bond enthalpies, as shown below:

$$CH_4 + 2O_2 \rightarrow CO_2 + 2H_2O$$

It is useful to draw out the reaction using displayed formulae, so that all the bonds that are broken and formed can be seen clearly.

Bonds broken (endo: +)	kJ mol^{-1}		Bonds formed (exo: –)	kJ mol^{-1}
4 (C–H) = 4 × +413	+1652		2 × (C=O) = 2 × –805	–1610
2 (O=O) = 2 × +498	+996		4 × (O–H) = 4 × –464	–1856
Total enthalpy needed to break all bonds	+2648		Total enthalpy released to form all bonds	–3466
Enthalpy change for the reaction ΔH= +2648 – 3466 = –818 kJ mol^{-1}				

The accepted value for this reaction is –890 kJ mol^{-1}, which differs substantially from the value calculated above. This can be explained largely by the fact that we use average bond enthalpies for the C–H, C=O and O–H bonds, and not specific bond enthalpies.

Enthalpy changes using Hess's law

Many chemical reactions cannot be measured directly. This can be for either energetic or kinetic reasons. If the activation energy is very high or if a reaction rate is very slow, then the enthalpy change cannot be measured directly by experiment. Reactions that cannot be measured directly can be calculated indirectly by using energy cycles. Energy cannot be created or destroyed and therefore the enthalpy change for a reaction is independent of the route taken.

Hess's law states that if a reaction can take place by more than one route, the overall enthalpy change for each route is the same, irrespective of the route taken, provided that the initial and final conditions are the same.

When applying Hess's law it is helpful to construct an enthalpy triangle.

It is difficult to measure the enthalpy of formation of CO(g) directly, but it can be calculated using Hess's law.

The enthalpy change for the following reactions can be measured experimentally:

$$C(s) + O_2(g) \rightarrow CO_2(g) \qquad \Delta H = -394\,\text{kJ mol}^{-1}$$

$$CO(g) + \tfrac{1}{2}O_2(g) \rightarrow CO_2(g) \qquad \Delta H = -284\,\text{kJ mol}^{-1}$$

It is always best to start with the enthalpy change that has to be calculated. You can then write that along the top and call it ΔH_1.

$$C(s) + \tfrac{1}{2}O_2(g) \rightarrow CO(g)$$

Then construct a cycle with two alternative routes (Figure 45).

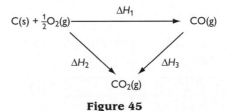

Figure 45

The simplest way to apply Hess's law is then to look at the direction of the arrows. Route 1 is made up of arrows that point in the clockwise direction, while route 2 consists of arrows that point in the anti-clockwise direction (Figure 46).

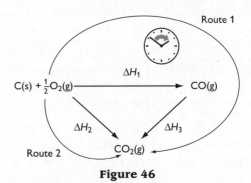

Figure 46

route 1 = route 2, therefore $\Delta H_1 + \Delta H_3 = \Delta H_2$

$$\Delta H_1 = \Delta H_2 - \Delta H_3$$
$$= -394 - (-284)$$
$$= -394 + 284$$
$$= -110\,\text{kJ}\,\text{mol}^{-1}$$

So, for $C(s) + \frac{1}{2}O_2(g) \rightarrow CO(g)$, $\Delta H = -110\,\text{kJ}\,\text{mol}^{-1}$.

There are two possible enthalpy cycles that you may be asked to construct. If you are asked to calculate the enthalpy change of formation and are given data on the enthalpy of combustion, it is best to construct a cycle like the one shown below, where the combustion products are shown at the bottom. The arrows always point downwards to the combustion products (Figure 47).

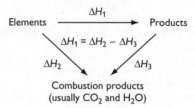

Figure 47

If Hess's law is applied to the cycle:

$\Delta H_1 + \Delta H_3 = \Delta H_2$ hence $\Delta H_1 = \Delta H_2 - \Delta H_3$

The alternative calculation is when you are asked to calculate the enthalpy change of combustion and you are given data on the enthalpy changes of formation. The elements are best shown at the bottom, and the arrows point up (Figure 48).

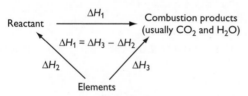

Figure 48

If Hess's law is applied to the cycle:

$\Delta H_1 + \Delta H_2 = \Delta H_3$ hence $\Delta H_1 = \Delta H_3 - \Delta H_2$

Example

Use the data in the table below to calculate the standard enthalpy change of combustion for ethane:

Substance	ΔH_f°/kJ mol^{-1}
$C_2H_6(g)$	−85
$CO_2(g)$	−394
$H_2O(l)$	−286

Step 1: write an equation for the enthalpy of combustion required:

$$C_2H_6(g) + 3\tfrac{1}{2}O_2(g) \rightarrow 2CO_2(g) + 3H_2O(l)$$

Step 2: construct the enthalpy triangle by writing the elements at the bottom. Both arrows will point upwards (Figure 49).

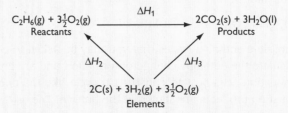

Figure 49

Step 3: apply Hess's law using the clockwise = anti-clockwise rule:

$$\Delta H_1 + \Delta H_2 = \Delta H_3 \qquad\qquad \text{hence } \Delta H_1 = \Delta H_3 - \Delta H_2$$

$\Delta H_2 = $ formation of $C_2H_6(g) = -85\,kJ\,mol^{-1}$

$\Delta H_3 = 2 \times$ [formation of $CO_2(g)$] + 3 × [formation of $H_2O(l)$]

$\qquad = (2 \times -394) + (3 \times -286)$

$\qquad = -1646\,kJ\,mol^{-1}$

therefore

$\Delta H_1 = (-1646) - (-85) = -1561\,kJ\,mol^{-1}$

$C_2H_6(g) + 3\tfrac{1}{2}O_2(g) \rightarrow 2CO_2(g) + 3H_2O(l) \qquad \Delta H_c^\circ = -1561\,kJ\,mol^{-1}$

Examiner tip

Most exam papers contain at least one question using Hess's law and there are only two types of question that can be asked. If the data provided is standard enthalpy of:

- combustion, the cycle will have combustion products at the bottom and the arrows will point down
- formation, the cycle will have elements at the bottom and the arrows will point up

Rates

Experimental observations show that the rate of a reaction is influenced by:
- temperature
- concentration
- the use of a catalyst

The collision theory of reactivity helps to provide explanations for these observations. A reaction cannot take place unless a collision occurs between the reacting particles. Increasing the temperature or concentration increases the chance that a collision will occur.

Knowledge check 22

Explain how increasing the pressure on a gaseous reaction affects the rate of reaction.

However, not all collisions lead to a successful reaction. The energy of a collision between reacting particles must exceed the minimum energy required to start the reaction. This minimum energy is known as the activation energy, E_a. Increasing the temperature affects the number of collisions that have an energy exceeding E_a, while the use of a catalyst lowers E_a.

Boltzmann distribution of molecular energies

Energy is directly proportional to absolute temperature. When collisions occur, the particles involved in the collision exchange (gain or lose) energy, even if a reaction does not occur. It follows that for any given mass of gaseous reactants, at a constant temperature, the distribution of energies will mean that some particles have more energy than others.

Figure 50 shows a typical distribution of energies at constant temperature, known as the Boltzmann distribution.

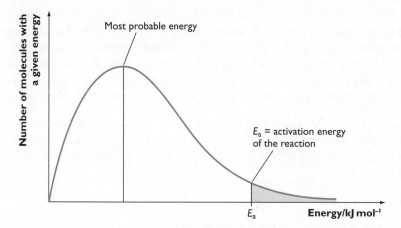

Figure 50

- The distribution always goes through the origin, showing that there are no particles with zero energy.
- The distribution is asymptotic (i.e. the curve approaches the axis, but will only meet the axis at infinity) to the horizontal axis at high energy. This shows that there is no maximum energy.
- E_a represents the activation energy, which is the minimum energy required to start the reaction.
- The area under the curve represents the total number of particles.
- The shaded area represents the number of particles with energy greater than or equal to the activation energy, $E \geq E_a$ (showing the number of particles with sufficient energy to react — Figure 51).

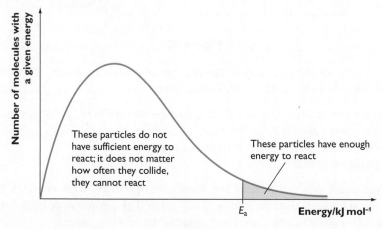

Figure 51

Effect of concentration, temperature and catalysts on the rate of reaction

Concentration

A useful analogy is to imagine your first driving lesson. The one thing that you want to avoid is a collision. It follows that your first lesson is likely to be early on a Sunday morning on a quiet country lane, rather than at 5.00 p.m. on a Friday evening in the city centre. It is obvious that the high concentration of cars at rush hour increases the chance of a collision. The same is true for a chemical reaction. Increasing concentration simply increases the chance of a collision. The more collisions there are, the faster the reaction will be.

For a gaseous reaction, increasing the pressure has the same effect as increasing the concentration. When gases react, they react faster at high pressure, because as the pressure increases so does the concentration, leading to an increased chance of a collision.

Temperature

Examiner tip

When drawing the Boltzmann distribution at a higher temperature the marking points are:

- the curve goes through the origin and there are fewer particles with low energy
- the most probable energy moves to the right (higher energy) but the height of the peak is lower
- there are more particles with high energy such that a greater proportion of particles have energy that exceeds the activation energy

An increase in temperature has a dramatic effect on the distribution of energies, as can be seen in Figure 52.

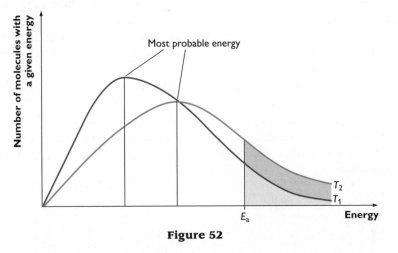

Figure 52

Only the temperature has changed. The number of particles remains the same, and therefore the area under both curves remains the same.

At higher temperature (T_2) the distribution flattens and shifts to the right, so that:
- there are fewer particles with low energy
- the most probable energy moves to higher energy
- more particles have $E \geq E_a$

Increasing temperature increases the number of particles with energies greater than or equal to the activation energy, $E \geq E_a$, so at high temperature there are more particles with sufficient energy to react, making the reaction faster.

Decreasing the temperature has the opposite effect.

OCR(A) AS Chemistry

Catalysts

From GCSE, you will recall that catalysts speed up reactions without themselves changing permanently. Catalysts work by lowering the activation energy for the reaction. This is illustrated below by an energy-profile diagram (Figure 53) and by the Boltzmann distribution on p. 49.

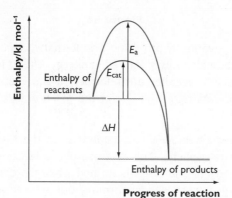

Figure 53

E_a is the activation energy of the uncatalysed reaction. E_{cat} is the activation energy of the catalysed reaction. E_{cat} is lower than E_a.

A catalyst lowers the activation energy but does not alter the Boltzmann distribution. It follows that more particles have $E \geq E_{cat}$.

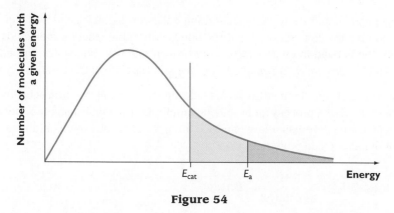

Figure 54

The shaded areas in Figure 54 indicate the proportion of particles whose energy exceeds the activation energies.

Homogeneous catalysts

The catalyst is in the same phase (gas, liquid or solid) as the reactants. The most common phase is a liquid, as many reactions are carried out in aqueous solution. A good example of a reaction involving a homogeneous catalyst is esterification.

The reaction involves ethanol (a liquid) and ethanoic acid (a liquid) and uses sulfuric acid (also a liquid) as the catalyst (Figure 55).

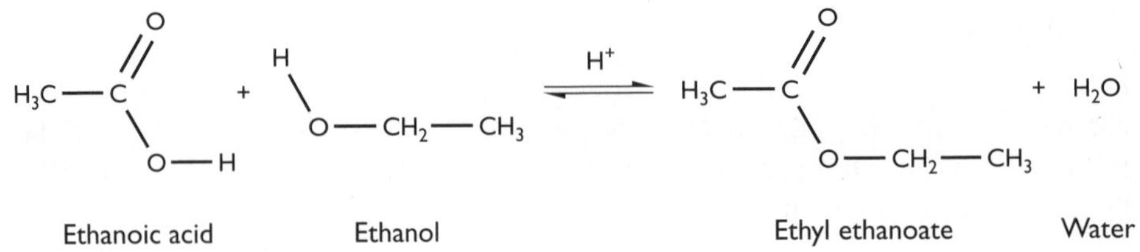

Ethanoic acid   Ethanol   Ethyl ethanoate   Water

Figure 55

The acid catalyst, H^+, works by providing an alternative mechanism involving an activated intermediate with a lower activation energy. The H^+ ions from the catalyst take part in the reaction, but they are reformed at the end. This means that the concentration of acid catalyst at the end of the reaction is the same as when the reagents were mixed.

A second example of a homogeneous catalyst is in the loss of ozone from the upper atmosphere (stratosphere). This is a gas-phase reaction and the catalyst is the chlorine radical, $Cl\bullet(g)$. The $Cl\bullet(g)$ is formed when the C–Cl bonds in CFCs are exposed to high-energy ultraviolet light, forming radicals. The reactions are complex, but can be summarised as shown on p. 34.

Heterogeneous catalysts

The catalyst is in a different phase to the reactants. The most common type of heterogeneous catalysis involves reactions of gases with a solid catalyst. Heterogeneous catalysts also lower the activation energy, but their mode of action is different to that of a homogeneous catalyst.

Heterogeneous catalysts work by adsorbing gases on to their solid surfaces. This adsorption results in a weakening of the bonds within the reactant molecules, thus lowering the activation energy. Bonds are broken and new bonds are formed. The product molecules are then desorbed from the solid surface of the catalyst.

Transition metals are frequently used as heterogeneous catalysts. Iron is used as the catalyst in the Haber process for producing ammonia. The iron is usually either finely divided (and therefore has a large surface area), or is porous and contains a small amount of metal oxide promoters.

$$N_2(g) + 3H_2(g) \xrightarrow[\text{as catalyst}]{\text{Fe(s)}} 2NH_3(g)$$

The motor car, with the internal combustion engine, is responsible for many of the pollutants discharged into the atmosphere. However, modern cars are now fitted with catalytic converters, which have contributed significantly to an improvement in air quality. Cars often discharge unreacted hydrocarbons into the atmosphere. Some of these, for example benzene, are toxic and carcinogenic.

Carbon monoxide is also formed by the incomplete combustion of fuel.

$$C_8H_{18} + 8\tfrac{1}{2}O_2 \rightarrow 8CO + 9H_2O$$

Examiner tip

If you are asked how a catalyst works, look at how many marks are available. If there is more than 1 mark, the explanation required is usually that the catalyst works by firstly adsorbing gases onto its surface ✔, which weakens the bonds and lowers activation energy and leads to a chemical reaction ✔ followed by the products of the reaction desorbing from the surface of the catalyst ✔.

OCR(A) AS Chemistry

The internal combustion engine often reaches temperatures of around 1000°C. These high temperatures provide sufficient energy for nitrogen and oxygen (both present in the air) to react and form oxides of nitrogen.

$$N_2 + O_2 \rightarrow 2NO$$

$$2NO + O_2 \rightarrow 2NO_2$$

Carbon monoxide, oxides of nitrogen (NO and NO_2) and unburnt hydrocarbons can lead to the formation of photochemical smog. Under certain conditions, such as bright sunlight and still air conditions, this can produce low-level ozone. High-level ozone, which occurs in the upper atmosphere, is beneficial, but low-level ozone traps pollutant gases.

The catalytic converter, fitted to all modern cars, reduces the emission of unburnt hydrocarbons, carbon monoxide and oxides of nitrogen. The equations below show how this is achieved.

Removal of unburnt hydrocarbons:

$$C_8H_{18}(g) + 12\frac{1}{2}O_2(g) \rightarrow 8CO_2(g) + 9H_2O(g)$$

Removal of carbon monoxide and oxides of nitrogen:

$$2NO(g) + 2CO(g) \rightarrow N_2(g) + 2CO_2(g)$$

$$2NO_2(g) + 4CO(g) \rightarrow N_2(g) + 4CO_2(g)$$

All of the above reactions occur in the gas phase. The catalytic converter consists of a fine aluminium mesh coated with a thin, solid film of an alloy of platinum, rhodium and palladium. Catalytic converters only function efficiently at high temperatures and are therefore not very effective on short journeys.

Catalysts are of great economic importance. The use of a catalyst often allows:
• reactions to be carried out at lower temperatures and pressures, thus reducing costs
• different reactions to be used, with lower atom economy and therefore with reduced waste

Enzymes, which operate close to room temperature and pressure, are increasingly used to generate a specific product.

Dynamic equilibrium and le Chatelier's principle

Reversible reactions

There are many everyday examples of reversible reactions or processes — the most common being the physical states of water. If the temperature of water falls below 0°C, the water will freeze and ice will form. However, when the temperature rises above 0°C, the ice melts and reforms the water. This process can be represented as:

$$H_2O(s) \rightleftharpoons H_2O(l)$$

The ⇌ sign indicates that a reaction is reversible. This type of reaction is common in many chemical reactions, such as esterification (Figure 56).

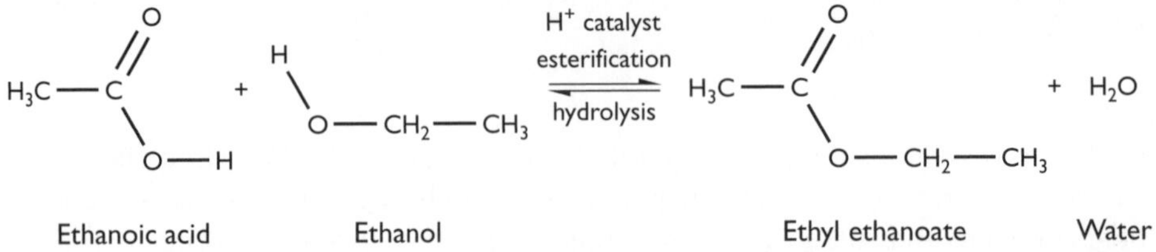

Ethanoic acid   Ethanol   Ethyl ethanoate   Water

Figure 56

Ethanoic acid reacts with ethanol, in the presence of an acid catalyst, to produce ethyl ethanoate and water. However, the ester, ethyl ethanoate, is hydrolysed by water, in the presence of an acid catalyst, and reforms ethanoic acid and ethanol.

Dynamic equilibrium

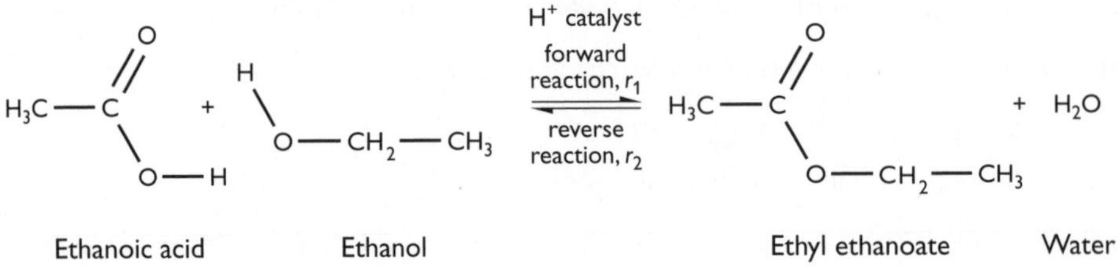

Ethanoic acid   Ethanol   Ethyl ethanoate   Water

Figure 57

In the reversible reaction shown in Figure 57, the reaction moving from left to right is known as the **forward reaction**, and the reaction moving from right to left is referred to as the **reverse reaction**.

The forward reaction r_1 starts fast but slows down, while the reverse reaction r_2 starts slowly and speeds up. It follows that there will be a point at which the rate of the forward reaction exactly equals the rate of the reverse reaction.

$$r_1 = r_2$$

When this happens, the system is said to be in dynamic equilibrium, i.e. **equilibrium** has been reached because the amount of each chemical in the system remains constant. It is **dynamic** because the reagents and the products are constantly interchanging. Dynamic equilibrium can only be reached if the system is closed.

A useful definition to learn for examinations is that a dynamic equilibrium is reached when the rate of the forward reaction equals the rate of the reverse reaction, such that the concentration of the reagents and products remains constant, while the reagents and the products constantly interchange.

Knowledge check 23

Explain what is meant by the terms *reversible reaction* and *dynamic equilibrium*.

Le Chatelier's principle

The French chemist Henri le Chatelier studied many dynamic equilibria. He suggested a general qualitative rule to predict the movement of the position of the equilibrium.

Le Chatelier's principle states that if a closed system under equilibrium is subject to a change, then the system will move in such a way as to minimise the effect of the change.

The factors that can be readily changed are concentration, temperature and pressure, and the use of a catalyst.

Effect of changing concentration on the position of the equilibrium

The equilibrium formed between the chromate ion, CrO_4^{2-}, and the dichromate ion, $Cr_2O_7^{2-}$, is useful because each ion is coloured, and it is therefore possible to observe the movement of the position of the equilibrium.

$$2H^+ \; + \; 2CrO_4^{2-} \; \rightleftharpoons \; Cr_2O_7^{2-} \; + \; H_2O$$
$$\;\;\;\;\;\;\;\;\; \text{yellow} \;\;\;\;\;\;\;\; \text{orange}$$

If we add some acid to the chromate/dichromate mixture, the concentration of the acid, $[H^+]$, increases. The system will now move in such a way as to minimise the effect, i.e. it will try to decrease the concentration of the acid, $[H^+]$. It can only achieve this if the additional H^+ reacts with some of the CrO_4^{2-} to form the products $Cr_2O_7^{2-}$ and H_2O. Put in very simple terms, by adding acid we add to the left-hand side. The system counters this by moving to the right-hand side, and the colour changes to orange.

Knowledge check 24

State le Chatelier's principle.

Effect of changing pressure on the position of the equilibrium

Pressure has virtually no effect on the reactions of solids or liquids and only affects gaseous reactions. The pressure of the gas mixture simply depends on the number of gas molecules in the mixture. The greater the number of gas molecules in the equilibrium mixture, the greater the pressure in the equilibrium mixture. If the pressure is increased, then the system under equilibrium will try to decrease the pressure by reducing the number of gas molecules in the system.

If the pressure is increased in a system such as $2SO_2(g) + O_2(g) \rightleftharpoons 2SO_3(g)$, the position of the equilibrium will move to the *right*, reducing the number of molecules. This has the effect of reducing the pressure.

If the pressure is increased in a system such as $N_2O_4(g) \rightleftharpoons 2NO_2(g)$, the position of the equilibrium will move to the *left*, reducing the number of molecules. This has the effect of reducing the pressure.

If the pressure is increased on a system such as $2HI(g) \rightleftharpoons H_2(g) + I_2(g)$, the position of the equilibrium will *not* move to the right or to the left, because there are the same number of molecules on each side of the equilibrium, and movement from one side to the other has no effect on the pressure.

Increasing pressure also increases the rate of reaction.

Effect of changing temperature on the position of the equilibrium

Temperature not only influences the rate of the reaction, but also plays an important role in determining the position of the equilibrium. The effect of temperature can only be predicted if the ΔH value of the reaction is known.

For the reaction:

$$2A(g) + B(g) \rightleftharpoons C(g) + D(g) \qquad \Delta H = -100\,kJ\,mol^{-1}$$

it follows that the forward reaction:

$$2A(g) + B(g) \rightarrow C(g) + D(g) \qquad \Delta H = -100\,kJ\,mol^{-1}$$

is exothermic, and the reverse reaction:

$$C(g) + D(g) \rightarrow 2A(g) + B(g) \qquad \Delta H = +100\,kJ\,mol^{-1}$$

is endothermic.

If we increase the temperature for the reaction mixture, then, according to le Chatelier's principle, the system will try to decrease the temperature. The equilibrium mixture can achieve this by favouring the reverse reaction, which is endothermic, and therefore removes the additional enthalpy caused by increasing the temperature. The position of the equilibrium moves to the left.

$2HI(g) \rightleftharpoons H_2(g) + I_2(g)$ is an *endothermic* reaction, and therefore if:
- the temperature is increased then the equilibrium moves to the right
- the temperature is decreased then the equilibrium moves to the left

$SO_2(g) + O_2(g) \rightleftharpoons 2SO_3(g)$ is an *exothermic* reaction, and therefore if:
- the temperature is increased then the equilibrium moves to the left
- the temperature is decreased then the equilibrium moves to the right

Effect of using a catalyst on the position of the equilibrium

The definition of a catalyst is a substance that speeds up the rate of reaction, without itself being changed, by lowering the activation energy of the reaction. In addition, the catalyst does not alter the amount of product produced.

In a system under equilibrium, a catalyst will speed up the forward and the reverse reactions equally, and will therefore have no effect on the position of the equilibrium. However, catalysts still play an important part in reversible reactions, as they reduce the time taken to reach equilibrium. In brief, a catalyst will produce the same amount of product, but will produce it more quickly.

The Haber process

We need large amounts of nitrogen compounds, particularly for fertiliser, and the Haber process is essential for the 'fixation' of atmospheric nitrogen. Atmospheric nitrogen is in plentiful supply, but cannot be used directly and must be converted (i.e. fixed) into a useful compound. The Haber process converts nitrogen into ammonia.

$$N_2(g) + 3H_2(g) \rightleftharpoons 2NH_3(g) \qquad\qquad \Delta H = -93\,kJ\,mol^{-1}$$

Le Chatelier's principle allows us to predict the optimum conditions for this industrial process.

ΔH is $-93\,kJ\,mol^{-1}$, which tells us that the forward reaction is exothermic and, therefore, if:
- the temperature is increased then the equilibrium moves to the left
- the temperature is decreased then the equilibrium moves to the right

The optimum temperature to achieve the maximum yield of ammonia is therefore a low temperature.

If the pressure is increased on a system such as $N_2(g) + 3H_2(g) \rightleftharpoons 2NH_3(g)$, then the position of the equilibrium will move to the *right*, so that the number of molecules is reduced, which has the effect of reducing the pressure.

The optimum pressure to achieve the maximum yield of ammonia is therefore a high pressure.

The table below illustrates the effect of changing temperature and pressure on the yield of ammonia.

Percentage yield of ammonia

Temperature/K	Pressure/atm			
	25	50	100	200
373	92	94	96	98
573	28	40	53	67
773	3	6	11	18
973	1	2	4	9

Low temperature (373 K = 100°C) gives the highest percentage yield. However, at low temperature the rate of reaction is slow, and a compromise has to be reached between yield and rate of reaction.

High pressure (200 atm) gives the highest percentage yield. High pressure also increases the rate of reaction. However, at high pressure the operating costs increase and a compromise has to be reached between yield, rate, safety and cost.

The maximum conversion rate is 98%, and the optimum conditions are 373 K and 200 atm.

Manufacturing conditions

The manufacture of ammonia in a modern plant is highly efficient. The operating conditions are a compromise:
- a temperature of around 700 K (427°C)
- a pressure of around 100 atm

The rate of reaction is greatly improved by using a finely divided or porous iron catalyst, which incorporates metal oxide promoters.

Summary

Having revised **Module 3: Enthalpy changes, rates and equilibria** you should now have an understanding of:

- exothermic and endothermic reactions
- enthalpies of reaction, formation and combustion
- bond enthalpies
- Hess's law and calculations using enthalpy cycles
- Boltzmann distributions and how temperature and catalysts affect rate
- catalysts
- le Chatelier's principle

Module 4: Resources

Chemistry of the air

The greenhouse effect

The Earth is mostly warmed by the energy transmitted from the sun. This consists largely of the wavelengths of visible light, but there is also some ultraviolet and some infrared radiation. The ultraviolet radiation is largely removed in the upper parts of the atmosphere (known as the stratosphere) by the ozone layer.

Once at the Earth's surface, the many chemical reactions that occur absorb and transmit this energy. Over the years, the overall surface temperature has remained more or less constant, because an equilibrium has been struck between the arriving and departing energy. The departing energy is almost wholly infrared radiation. Many of the gases in the atmosphere absorb some of the infrared energy and reflect it back to the Earth. In other words, these gases act as a kind of thermal blanket. Carbon dioxide and water molecules are examples of gases that absorb infrared radiation.

Of the gases naturally present in the atmosphere, neither nitrogen nor oxygen absorb infrared radiation, but others do, and they are collectively known as 'greenhouse gases'. A wide range of gases contribute to the overall 'greenhouse effect' and the contribution of an individual gas to the greenhouse effect depends on:
- its ability to absorb infrared radiation
- its atmospheric concentration
- its residence time — how long it stays in the atmosphere

You should be aware that the C=O bonds in carbon dioxide, the C–H bonds in methane, and the O–H bonds in water all absorb infrared radiation of a particular wavelength by vibrating with increased energy. This energy is then randomly dispersed, with much of it returning to the Earth's surface. It is inevitable that if the concentration of these gases is allowed to rise, the average temperature at the Earth's surface will increase.

Examiner tip

All chemistry exams have to incorporate different aspects of 'How Science Works' and over recent years this area of the specification has been examined extensively. It is important that you take time to learn all the relevant detail. Just because things like the greenhouse effect and global warming appear almost daily in the news, don't assume that you know it all. Look at the mark schemes and see exactly what the examiner expects you to write.

Much current attention has centred on the role of carbon dioxide as a contributor to global warming. However, carbon dioxide is not a notably efficient greenhouse gas, but it is, of course, produced in huge quantities by the burning of fossil fuels such as wood, coal and oil products; and this is something that should be within our power to control. (In fact, water vapour is the biggest contributor to the greenhouse effect.) Many attempts have been made to predict the result of continuing to release carbon dioxide into the atmosphere at the current rate. There is no universal agreement, but there is a serious concern that the melting of the polar ice cap could cause extensive flooding of low-lying land, and that changing patterns of temperature might lead to severe droughts in some parts of the world.

Scientists have an essential role to play and they must:
- Try to extend our knowledge of the processes involved, to allow predictions to be made with the greatest possible certainty.
- Make their findings known as widely as possible, so that governments can understand the impact that current findings may have in the future.
- Use initiatives, such as the Kyoto Protocol, to monitor the progress that is being made to provide a cleaner environment.
- Research alternative methods of providing sources of power and processes to manufacture chemicals with less environmental impact.
- Search for solutions to resolve current pollution issues.

> **Knowledge check 25**
>
> State what happens to the bonds in a CO_2 molecule when it absorbs infrared.

> **Knowledge check 26**
>
> State *three* factors that influence a gas's contribution to the greenhouse effect.

Carbon dioxide removal

Under certain conditions it is possible to liquefy CO_2. Scientists are currently investigating the possibility of capturing atmospheric CO_2 gas and converting it into liquid CO_2. This is gaining widespread approval and is known as carbon capture and storage (CCS). The liquid CO_2 can then be injected deep under the oceans for storage. Possible sites include depleted oil and gas fields. The obvious risk of the method is subsequent leakage of the gas, but in trials so far it has not proved to be a problem and it has been used in Norway for a number of years.

An alternative approach could be to react the carbon dioxide with metal oxides to form carbonates. However, a practical way of doing this on a large scale has not yet been devised.

> **Knowledge check 27**
>
> State *two* ways in which $CO_2(g)$ might be removed from the atmosphere.

The ozone layer

The ozone layer exists about some 25 km above the Earth's surface in the stratosphere. It has a depth of approximately 15 km. It absorbs much of the harmful ultraviolet radiation emitted by the sun, which, if it were to reach the Earth's surface unimpeded, would cause a considerable increase in burning and skin cancers among humans.

Ozone, O_3, is being formed continuously and broken down in the stratosphere, so that an equilibrium is established with molecular oxygen, O_2, and an oxygen radical, O.

$$O_3 \rightleftharpoons O_2 + O$$

It was shocking to discover that the ozone layer above the Antarctic during the polar spring season had thinned to the point where a hole had developed. The main culprits were almost certainly chlorine radicals, created by the homolytic breakage

> **Knowledge check 28**
>
> Explain how ozone is formed.

of a covalent bond. They can attack the ozone layer, as shown in the reaction scheme below.

First, there is an initiation step, resulting in the formation of chlorine radicals ($Cl\bullet$). This is most likely to be the result of the breaking of a C–Cl bond.

$$R\text{–}Cl \rightarrow R\bullet + Cl\bullet$$

The chlorine radicals are then involved in the propagation steps:

$$Cl\bullet + O_3 \rightleftharpoons ClO\bullet + O_2$$
$$ClO\bullet + O \rightleftharpoons Cl\bullet + O_2$$

You will notice that in the last step the chlorine radicals have been regenerated and can therefore recirculate, causing the destruction of many ozone molecules. It has been estimated that around 1000 ozone molecules may be destroyed as the result of the production of a single chlorine radical. The sequence is only terminated if the chlorine radical combines with another radical and is removed from the propagation sequence.

The main source of chlorine-containing chemicals is CFCs. However, other chlorine compounds are also involved, as many have been extensively used as solvents in processes such as dry cleaning.

Compounds containing a C–Cl bond are not alone in supplying a radical able to initiate the destruction of the ozone layer. Of growing concern is the role of nitrogen monoxide. A major source of this gas occurs during thunderstorms, when lightning causes nitrogen and oxygen to combine:

$$N_2(g) + O_2(g) \rightarrow 2NO(g)$$

Nitrogen monoxide is also present in the exhaust gases of aircraft.

A small amount of nitrogen monoxide will migrate to the stratosphere, where it is able to attack the ozone layer in a series of reactions analogous to those of the chlorine radical. No initiation step is required because nitrogen monoxide naturally possesses an unpaired electron. The propagation steps are as follows:

$$NO + O_3 \rightleftharpoons NO_2 + O_2$$
$$NO_2 + O \rightleftharpoons NO + O_2$$

Comparison with the propagation steps for the chlorine radical shows that these steps can be expressed more generally using the letter R to represent the radical:

$$R + O_3 \rightleftharpoons RO + O_2$$
$$RO + O \rightleftharpoons R + O_2$$

Controlling air pollution

Global warming and the damage to the ozone layer represent complex long-term problems that will be with us for decades to come. Other forms of atmospheric pollution are more direct and immediate. The gases emitted as by-products of industrial processes, the treatment of waste and the burning of fuels are often directly

---

Knowledge check 29

Write an equation to show how ultraviolet light could generate a $Cl\bullet$ radical from the CFC difluorochloromethane.

Knowledge check 30

Write equations to show how NO radicals can destroy ozone molecules.

toxic. They can sometimes take part in complex reactions, resulting in the build-up of other undesirable pollutants.

The internal combustion engine is responsible for many of the pollutants discharged into the atmosphere, although the use of catalytic converters has contributed significantly to reducing the level of the more immediately dangerous pollutants.

Without a catalytic converter, a car would discharge a number of gases into the atmosphere (see pp. 52–53).

Other air pollutants

As well as the gaseous products already mentioned, we have to take into account other types of pollutant, for example:

- Vehicles burning diesel may emit very fine particles of carbon in their exhausts. These particles are derived from the complete breakdown of the fuel. They can create breathing problems for asthmatics.
- Unburnt fuel, released into the atmosphere together with CO, NO, NO_2 and water vapour, forms a mixture that can undergo radical reactions. These reactions lead to the formation of ozone, which is toxic even at low concentrations.
- A further series of reactions can then take place, creating unexpected organic molecules such as $CH_3COO_2NO_2$ (known by its old name of peroxyacetylnitrate or PAN). This an unpleasant compound that irritates the mucous membrane and can cause severe breathing difficulties. Fortunately, PAN is usually only formed in the upper atmosphere. However, occasionally, conditions are such that mixing of the atmosphere occurs, and this molecule — along with nitrogen dioxide and ozone — occurs at sufficiently low altitudes to become a severe risk. The conditions required are bright sunlight, still air and a relatively high concentration of exhaust pollutants, such as might occur, for example, in the morning rush hour in a big city. The unpleasant cocktail of gases is referred to as **photochemical smog**, and it is characterised by a brown haze caused by the presence of higher than usual concentrations of coloured nitrogen dioxide molecules.

Air pollution was a significant problem throughout the Beijing Olympic Games (2008).

Green chemistry

The growing public awareness that the future well-being of our planet may depend on our ability to control the extent to which we pollute our environment has led to the establishment of a movement that is often referred to as 'green chemistry'. There are some clear principles that responsible scientists should follow, including:

- Industrial processes should minimise the production and use of hazardous chemicals.
- Processes should use renewable materials.
- Energy requirements should be minimised, if necessary by the use of selective catalysts and sustainable energy sources.
- Products should be biodegradable.
- Processes should achieve a high atom economy to avoid the unnecessary waste of materials.

Knowledge check 31

Write equations to show how $CO(g)$, $NO(g)$ and $NO_2(g)$ are removed from the exhaust gas of a car.

Knowledge check 32

Write an equation to show the formation of $NO(g)$ in a car engine.

Most of these principles may seem to be common sense, but they have not always been properly considered in the past.

Use of lead

A compound of lead called lead tetraethyl, $Pb(C_2H_5)_4$, used to be added to petrol for cars in order to ensure smoother combustion of the fuel. However, the toxic nature of the lead released through the exhaust led scientists to modify the composition of the petrol, making the addition of lead tetraethyl unnecessary. Lead has also been largely eliminated from paint and electrical components. Lead piping is still found in older houses, but is steadily being replaced by copper or plastic.

Of course, fossil fuels themselves are a non-renewable commodity and their replacement by renewable fuels such as ethanol and biodiesel is the subject of active research and debate.

Dry cleaning

Dry cleaning involves the use of a solvent other than water. Chlorinated solvents are often used, but they can have a detrimental effect on the environment. Tetrachloromethane, CCl_4, was used originally, but was abandoned when found to cause liver and kidney damage. Other chlorinated liquids have replaced it, and currently tetrachloroethene ($Cl_2C=CCl_2$) is often used. However, using any chlorinated chemical is undesirable, since it is difficult to dispose of, and inevitably it will escape into the atmosphere.

An alternative is to use liquid 'supercritical' carbon dioxide at temperatures of around 40°C and pressures between approximately 75 and 200 times normal atmospheric conditions. This liquid is an efficient cleaner, and the excess carbon dioxide can be readily absorbed by alkalis.

Supercritical carbon dioxide can be used in the foaming of polymers and as a solvent to extract caffeine from coffee beans or tea leaves, creating decaffeinated drinks.

Fire extinguishers

Some fires can be extinguished using water, but others require the use of halogenated organic liquids (CFCs), as they are not flammable. Two examples are Halon 1211, $CBrClF_2$, and Halon 1301, $CBrF_3$. However, under the Montreal Protocol, the use of both these chemicals has been banned since 1996.

One alternative is to use a mixture known as PYROCOOL FEF. This is a foaming agent that smothers the fire, putting it out. PYROCOOL FEF has other advantages: the bubbles that it forms collapse quite quickly and coat the surface on which it is sprayed, and it is both non-toxic and biodegradable. A more subtle point is that it contains chemicals that absorb the high energy from the fire and re-emit it at longer, less dangerous wavelengths, helping to stop the fire from spreading.

International action

Global pollution is one of an increasing number of issues that affects everyone, and there has to be sufficient political will to attempt a solution. A number of international agreements have been made that, although far from perfect, represent some progress. However, unfortunately, some countries have not always been prepared to accept their full conditions.

The Montreal Protocol on substances that deplete the ozone layer — agreed and signed by the EEC and other countries in 1987 — was an important step towards the phasing out of CFCs, halons and other chlorine-containing chemicals.

The Kyoto Protocol has led to targets for the reduction of greenhouse gases, which, even if they are proving difficult to meet, is at least a step in the right direction.

In 2001, a convention aimed at reducing or eliminating the use of certain organic chemicals was agreed by a range of countries. These are substances known as persistent organic pollutants (POPs), which are not easily degraded and can travel through the environment and become incorporated in the food chain. They include insecticides (e.g. DDT), compounds unintentionally produced during combustion (e.g. dioxins) and some components of paints and plastics.

A more general agreement was made by the United Nations in 1992 in Rio de Janeiro, which affirms the importance of cooperation between nations to address 'environmental human rights'. It is, of course, a serious issue that those countries producing airborne pollution may not be the ones that suffer from its effects. The Rio Declaration was intended to establish principles that provide a basic framework to resolve some of the conflicts that might arise.

It is increasingly the case that scientists must accept that much of their work is not purely concerned with solving problems, but may carry additional responsibilities to the society in which they work, and often to a much wider community.

Having revised **Module 4: Resources** you should now have an understanding of:

- the greenhouse effect and the capture and storage of CO_2
- the ozone layer and the effect of radicals such as Cl• and NO
- controlling air pollution and the effect of the catalytic converter

- the *five* principles of sustainability:
  (1) try to avoid using hazardous chemicals
  (2) use reactions that have high atomy economy
  (3) use renewable resources
  (4) use alternative energy sources
  (5) ensure waste products are either non-toxic or are recycled safely

All of this section relates to the environment and very clearly meets the need to examine How Science Works.

Summary

Questions & Answers

Approaching the unit test

Terms used in the unit test

You will be asked precise questions in the unit test, so you can save a lot of valuable time — as well as ensuring you score as many marks as possible — by knowing what is expected. Terms used most commonly are explained below.

Define

This requires a precise statement to explain a chemical term. It could involve specific amounts or conditions such as temperature and pressure.

Explain

This normally implies that a definition should be given, together with some relevant comment on the significance or context of the term(s) concerned, especially where two or more terms are included in the question. The amount of supplementary comment should be determined by the mark allocation.

State

This implies a concise answer with little or no supporting argument.

Describe

This requires you to state in words (but using diagrams where appropriate) the main points of the topic. It is often used with reference either to particular phenomena or to particular experiments. In the former instance, the term usually implies that the answer should include reference to observations associated with the phenomena. The amount of description should be determined by the mark allocation. You are not expected to explain the phenomena or experiments, but merely to describe them.

Deduce or predict

This means that you are not expected to produce the answer by recall but by making a logical connection between other pieces of information. Such information may be wholly given in the question or could depend on answers given in an earlier part of the question. 'Predict' implies a concise answer, with no supporting statement.

Outline

This implies brevity, i.e. restricting the answer to essential detail only.

Suggest

This is used in two contexts. It implies either that there is no unique answer or that you are expected to apply your knowledge to a novel situation that may not be formally in the specification.

Calculate

This is used when a numerical answer is required. Working should be shown.

Sketch

When this is applied to diagrams, it means that a simple, freehand drawing is acceptable. Nevertheless, care should be taken over proportions, and important details should be labelled clearly.

About this section

This section contains questions similar in style to those you can expect to see in your Unit F322: Chains, Energy and Resources examination. The limited number of questions means that it is impossible to cover all the topics and all of the question styles, but they should give you a flavour of what to expect. The responses that are shown are real students' answers to the questions.

There are several ways of using this section. You could:

- hide the answers to each question and try the question yourself. It needn't be a memory test — use your notes to see if you can actually make all the points that you ought to make
- check your answers against the students' responses and make an estimate of the likely standard of your response to each question
- check your answers against the examiner's comments to see if you can appreciate where you might have lost marks
- take on the role of the examiner and mark each of the student's responses and then check to see if you agree with the marks awarded by the examiner

The examination lasts 105 minutes and there is a total of 100 marks. As you can imagine, time is tight. It is important that you practise answering questions under timed conditions.

Each question in this section identifies the specification topic, the total marks and a suggested time that should be spent writing out the answer.

Examiner comments

Examiner comments on the questions are preceded by the icon ⓔ. They offer tips on what you need to do in order to gain full marks. All student responses are followed by examiner's comments, indicated by the icon ⓔ, which highlight where credit is due. In the weaker answers, they also point out areas for improvement; specific problems; and common errors such as lack of clarity, irrelevance, misinterpretation of the question and mistaken meanings of terms.

Question 1 Alkanes and halogenoalkanes

Time allocation: 6–7 minutes

(a) Ethane, C_2H_6, reacts with chlorine (Cl_2) in the presence of sunlight to form a mixture of chlorinated products. One possible product is $C_2H_4Cl_2$.
   (i) State the type of mechanism involved in this reaction.   (1 mark)
   (ii) The initiation step involves the homolytic fission of the Cl–Cl bond. What is meant by the term homolytic fission?   (1 mark)

(b) Name the two possible isomers of $C_2H_4Cl_2$.   (2 marks)

(c) When $C_2H_4Cl_2$ is treated with aqueous NaOH, it undergoes substitution reactions to form both $C_2H_4(OH)Cl$ and $C_2H_4(OH)_2$.
   (i) State, and explain, the role of the $OH^-(aq)$ in these reactions.   (2 marks)
   (ii) Draw two possible isomers of $C_2H_4(OH)_2$.   (2 marks)

Total: 8 marks

🅔 The command word 'state' used in part (a) (i) indicates that a brief answer is required with no supporting argument. In (c) 'state and explain' requires a simple standard definition and also an explanation of the initial statement. In (b) 'name' means what it says. Oddly a substantial number of students ignore the instructions and draw the structures of the two isomers.

Student A

(a) **(i)** Radical substitution

Student B

(a) **(i)** Radical

🅔 Student A gets 1 mark, but student B scores no marks. The key word is 'substitution'.

Student A

(a) **(ii)** The bond breaks (i.e. fission), so that each atom in the bond retains one of the bonded electrons. For example:

    $Cl–Cl \rightarrow 2Cl\bullet$

Student B

(a) **(ii)** Radicals are produced by homolytic fission.

🅔 Student A gives a perfect answer and scores 1 mark, while student B scores no marks. Student B has missed the point of the question and has stated a consequence of homolytic fission rather than explaining what is *meant* by homolytic fission.

Student A

(b) 1,2-dichloroethane and 1,1-dichloroethane

Student B

(b) 1,1-dichloroethane and 1,2-dichloroethane

ⓔ Both answers are correct, for 2 marks.

Student A

(c) (i) Nucleophile, because it donates an electron pair to the $C^{\delta+}$.

Student B

(c) (i) OH⁻ is a nucleophile.

ⓔ Student A gets 2 marks, but student B only scores 1 mark for *stating* that it is a nucleophile. The clue is in the question, and a simple *explanation* is required for the second mark.

Student A

(c) (ii)

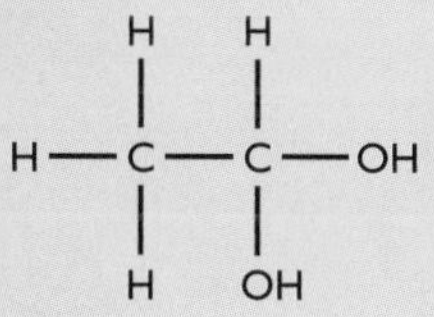

Student B

(c) (ii)

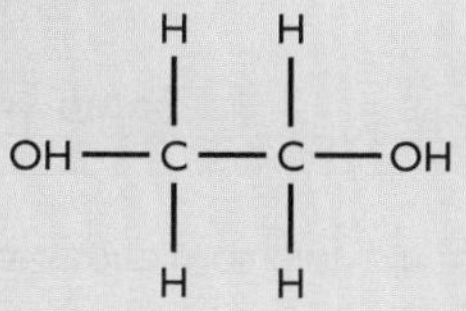

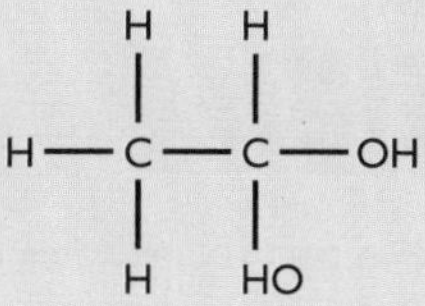

ⓔ Student A scores only 1 mark, because he/she has not read the instructions carefully and has drawn only one isomer. Student B scores 1 mark, but could have lost both marks. Student B recognises that it is possible to form ethane-1,2-diol and ethane-1,1-diol (although the latter doesn't exist), but then loses a mark by carelessly drawing the C–OH bond to the H and not to the O, i.e. C–HO. This is known as a bond-linkage error and is usually only penalised once in each paper.

The isomers should have been drawn as:

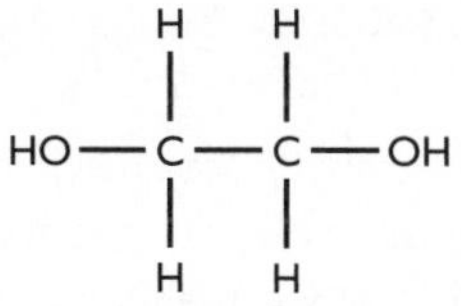

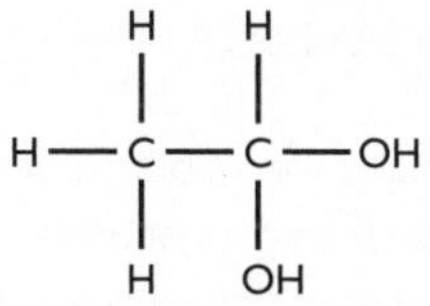

ⓔ **Student A does well and scores 7 marks out of 8, but carelessly loses a mark in the final part. Student B only gains 4 out of 8 marks. Student B's overall response places him or her as borderline grade D, but with a little care this could have been pushed up by 2 or 3 marks, to the A/B borderline.**

Question 2 Alcohols

Time allocation: 9–10 minutes

Compound A, shown below, contains two functional groups.

$$H-\overset{\overset{\displaystyle H}{|}}{\underset{\underset{\displaystyle H}{|}}{C}}-\overset{\overset{\displaystyle O}{||}}{C}-\overset{\overset{\displaystyle H}{|}}{\underset{\underset{\displaystyle H}{|}}{C}}-OH$$

Compound A

(a) Name the functional groups. (3 marks)

(b) Deduce the molecular formula of compound A. (1 mark)

(c) Compound A can be oxidised to produce a mixture of compound B, molecular formula $C_3H_4O_2$, and compound C, molecular formula $C_3H_4O_3$.
(i) Identify which of the functional groups could be oxidised. (1 mark)
(ii) State a suitable oxidising mixture. (2 marks)
(iii) State what you would observe during the oxidation. (1 mark)
(iv) Identify compound B. (1 mark)
(v) Write a balanced equation for the formation of compound C from compound A.
    Use [O] to represent the oxidising agent. (2 marks)

Total: 11 marks

ⓔ The command word 'deduce' in part (b) indicates that you should use information given in the question. By contrast, 'identify' in part (c) (iv) allows you flexibility — you could either name compound B or identify it by its structural, displayed or skeletal formula. The identification must be unambiguous, so the molecular formula would not score the mark.

Student A

(a) Primary alcohol and a ketone

Student B

(a) Ketone and an alcohol

ⓔ Student A scores all 3 marks but student B drops 1 mark by not *classifying* the alcohol. It is important that you use the information in the question. Each part of a question will tell you how many marks are allocated to that part, and it is essential that you use this information. Here there are 3 marks available, so three points are needed: ketone ✔, primary ✔ and alcohol ✔.

Student A

(b) $C_3H_6O_2$

Student B

(b) CH_3COCH_2OH

OCR(A) AS Chemistry

ⓔ Student A gets the mark, but student B has given a *structural* formula rather than a *molecular* formula. Molecular formulae always group together all atoms of the same element, and they are always written in the form: $C_xH_yO_z$. Student B gains no marks.

Student A

(c) (i) Alcohol

Student B

(c) (i) Alcohol

ⓔ Each student gains the mark.

Student A

(c) (ii) Acidified dicromate

Student B

(c) (ii) $H^+/Cr_2O_7^-$

ⓔ Student A gets both marks, while student B loses a mark by using an incorrect formula for the dichromate ion. Student A spelt dichromate incorrectly, but it is highly unlikely that this would be penalised. Spelling is only penalised if it states in the question that marks have been allocated for quality of written communication (QWC). However, the same tolerance is not extended to incorrect formulae. The formula should be $Cr_2O_7^{2-}$; student B didn't include the correct charge on the ion and therefore loses the mark.

Student A

(c) (iii) Colour change

Student B

(c) (iii) Turns green

ⓔ Neither student scores the mark. Student A's response is too vague. Student B hasn't stated the full colour change from orange to green.

Student A

(c) (iv)

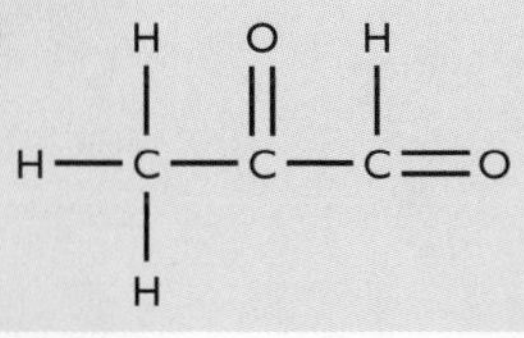

Student B

(c) (iv) B is CH_3COCOH

ⓔ Student A gains the mark, but student B doesn't score. Student A has made good use of the information in the question and used the displayed formula of compound A to deduce the structure of B. Student B's response is also good, but perhaps a little too clever. The final part of

CH_3COCOH indicates an alcohol. If it had been written as CH_3COCHO, it would have scored the mark. Student B clearly has potential but needs to take more care.

Student A

(c) (v)

$$H-\overset{\overset{\displaystyle H}{|}}{\underset{\underset{\displaystyle H}{|}}{C}}-\overset{\overset{\displaystyle O}{\|}}{C}-\overset{\overset{\displaystyle H}{|}}{\underset{\underset{\displaystyle H}{|}}{C}}-OH \ + \ [O] \ \longrightarrow \ H-\overset{\overset{\displaystyle H}{|}}{\underset{\underset{\displaystyle H}{|}}{C}}-\overset{\overset{\displaystyle O}{\|}}{C}-\overset{\overset{\displaystyle OH}{|}}{C}=O \ + \ H_2$$

Student B

(c) (v) $C_3H_6O_2 + 2[O] \rightarrow C_3H_4O_3 + H_2O$

ⓔ The correct answer is:

$$H-\overset{\overset{\displaystyle H}{|}}{\underset{\underset{\displaystyle H}{|}}{C}}-\overset{\overset{\displaystyle O}{\|}}{C}-\overset{\overset{\displaystyle H}{|}}{\underset{\underset{\displaystyle H}{|}}{C}}-OH \ + \ 2[O] \ \longrightarrow \ H-\overset{\overset{\displaystyle H}{|}}{\underset{\underset{\displaystyle H}{|}}{C}}-\overset{\overset{\displaystyle O}{\|}}{C}-\overset{\overset{\displaystyle OH}{|}}{C}=O \ + \ H_2O$$

The first mark would be awarded for the correct organic product and the second mark for balancing the equation.

Both students would score 1 out of 2 marks. Student A is very careful and makes good use of information in the question, but H_2 is never formed in the presence of an oxidising agent; water will always be produced. Student B is trying to be too clever. The equation is essentially correct and balanced, but, by using molecular formulae, the product hasn't been correctly identified. The formula $C_3H_4O_3$ is ambiguous and could be any one of several isomers.

ⓔ **Student A scores 9 out of 11 marks (which is grade-A standard) and student B would earn a grade E with 5 out of 11 marks.**

Question 3 **Fuels and isomerism**

Time allocation: 12–14 minutes

The hydrocarbons in crude oil can be separated by fractional distillation.

(a) Explain what is meant by the terms:
   (i) hydrocarbon (1 mark)
   (ii) fractional distillation (1 mark)

(b) Dodecane, $C_{12}H_{26}$, can be isolated by fractional distillation. Calculate the percentage composition by mass of carbon in dodecane. (3 marks)

(c) Dodecane can be cracked into octane and ethene only.
   Write a balanced equation for this reaction. (2 marks)

OCR(A) AS Chemistry

(d) Isomerisation of octane produces a mixture of isomers.

   (i) Name the isomers of C_8H_{18}.   (3 marks)

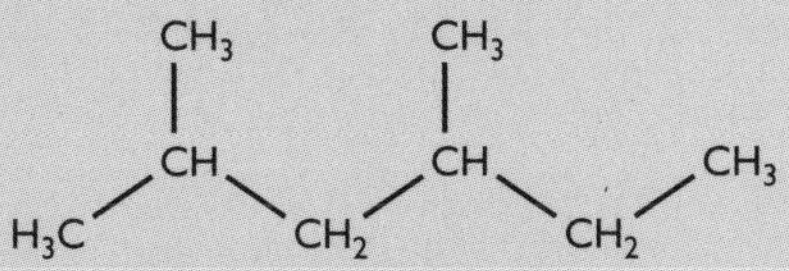

   (ii) Isomers A, B and C, shown above, can be separated by fractional distillation. State the order, lowest boiling point first, in which they would distil.   (1 mark)

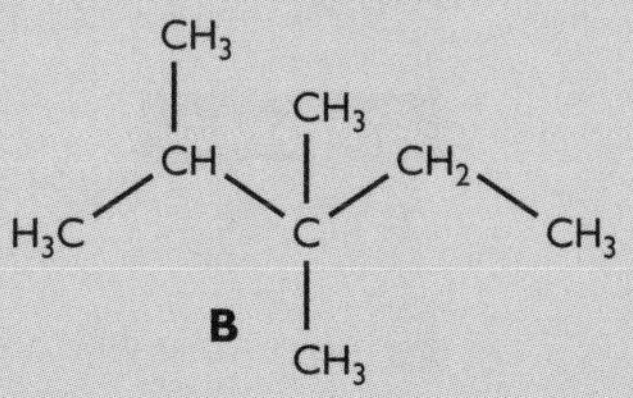

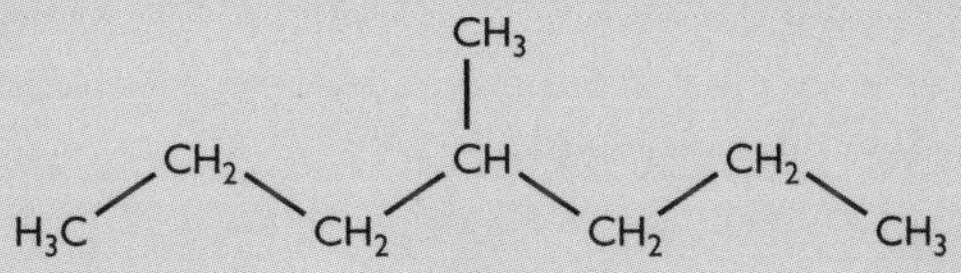

   (iii) Justify the order stated in (d)(ii).   (2 marks)

   (iv) Write a balanced equation for the *complete* combustion of octane, C_8H_{18}. (2 marks)

   (v) Why do oil companies isomerise alkanes such as octane?   (1 mark)

Total: 16 marks

ⓔ The command word 'calculate' used in (b) obviously requires a numerical approach. If more than 1 mark is allocated it is essential to show your working as any errors in the calculation will be marked consequentially. Incorrect answers may score some marks but only if you show your working. The command word 'justify' in (d) (iii) requires an explanation of your answer to (d) (ii). If your answer to (d) (ii) is incorrect, your justification will be marked consequentially.

Student A

(a) (i) C and H only

Student B

(a) (i) A hydrocarbon is a compound that contains carbon and hydrogen.

ⓔ Student A gains the mark, but student B does not. The key word missing in student B's response is *only*. Lots of compounds, such as ethanol (C_2H_5OH), contain carbon and hydrogen, but they are not hydrocarbons.

Student A

(a) (ii) Separation due to differences in boiling point

Student B

(a) (ii) Liquids with different boiling points are separated by fractional distillation.

ⓔ Both students gain the mark.

Student A

(b) $C_{12}H_{26}$ = 144 + 26 = 170
%C = (144/170) × 100 = 84.7%

Student B

(b) 84.7%

ⓔ Both students score 3 marks for the correct answer, but student A displays better exam technique by showing the working. If student B had made a mistake, then he/she would have scored zero, whereas if A had made a mistake, credit would have been given for the method. *Remember:* always show your working.

Student A

(c) $C_{12}H_{26} \rightarrow C_8H_{18} + 2C_2H_4$

Student B

(c) $C_{12}H_{26} \rightarrow C_8H_{18} + C_2H_4$

ⓔ Student A scores both marks. Student B gains only 1 mark for the correct formulae of octane and ethene. Student B lost the other mark because the equation wasn't balanced.

Student A

(d) (i) A = 2,4-dimethylhexane ✔
　　　 B = 2-dimethyl-4-methylpentane
　　　 C = 4-methylheptane ✔

Student B

(d) (i) A = 2,3-dimethylpentane
　　　 B = 2,3,3-trimethylpentane ✔
　　　 C = 3-methylheptane

ⓔ Student A gains 2 marks for A and C, while student B only scores 1 mark for B. Student B seems to have lost marks by carelessness, either when counting the length of the carbon chain or when counting the position of the side chain.

Student A

(d) (ii) BAC

Student B

(d) (ii) BAC

ⓔ Correct, for 1 mark.

Student A

(d) (iii) The longer the carbon chain, the more van der Waals forces there are and hence the higher the boiling point.

Student B

(d) (iii) Branched chains have lower boiling points.

ⓔ Student A gains both marks. A structural feature has been related to the trend in boiling point, and boiling point has been related to the amount of intermolecular bonding. Student B only scores 1 mark.

Student A

(d) (iv) $2C_8H_{18} + 25O_2 \rightarrow 16CO_2 + 18H_2O$

Student B

(d) (iv) $C_8H_{18} + 12\frac{1}{2}O_2 \rightarrow 8CO_2 + 9H_2O$

ⓔ Both students score 2 marks. It is acceptable to have fractions in the balanced equation.

Student A

(d) (v) Smaller chains are more useful.

Student B

(d) (v) Used as additives in petrol to increase the octane rating.

ⓔ Student B gives a good answer and gains the mark. Student A has confused 'isomerisation' with 'cracking', and scores no marks. Saying that smaller chains are more 'useful' is too vague to score the mark.

ⓔ **Student A scores a total of 14 out of 16 marks (88%), while student B scores 11 out 16 marks (69%). It is worth remembering that as a general guide: 80% = grade A, 70% = grade B, 60% = grade C etc.**

Question 4 **Radical and nucleophilic substitution**

Time allocation: 15–16 minutes

Ethane, C_2H_6, reacts with Cl_2 in the presence of sunlight to form a mixture of chlorinated products. One possible product is C_2H_5Cl, formed as shown in the following equation:

$$C_2H_6 + Cl_2 \rightarrow C_2H_5Cl + HCl$$

(a) Describe, with the aid of equations, the mechanism of this reaction. (4 marks)

(b) One other possible product of the reaction between ethane and chlorine is compound A, shown below.

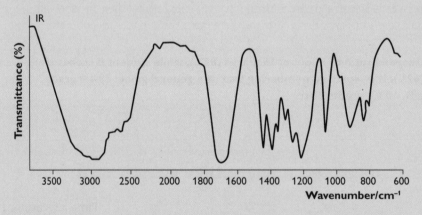

(i) **Name compound A.** (1 mark)
(ii) **Draw an isomer of compound A.** (1 mark)

(c) Chloroethane can react with a solution of sodium hydroxide, as follows:

$$C_2H_5Cl + OH^- \rightarrow C_2H_5OH + Cl^-$$

(i) **State the solvent in which the sodium hydroxide is dissolved.** (1 mark)
(ii) **State and explain the role of the hydroxide ion (OH⁻) in this reaction.** (2 marks)

(d) Ethanol, C_2H_5OH, is refluxed with an acidified solution of potassium dichromate(VI) to produce ethanoic acid. The acidified potassium dichromate(VI) acts as an oxidising agent.
(i) **Explain what is meant by the term *reflux*.** (1 mark)
(ii) **State what colour changes take place in the reaction mixture.** (1 mark)
(iii) **Write a balanced equation for the oxidation of ethanol to ethanoic acid. The oxidising agent can be represented as [O] in your equation.** (2 marks)

(e) An infrared spectrum of propanoic acid was obtained. By referring to your data sheet, identify two peaks in the infrared spectrum that confirm the presence of the carboxylic acid functional group.

(4 marks)

Total: 17 marks

ⓔ The command word 'describe' requires significant detail. When referring to a mechanism, equations are required and the movement of electrons must be tracked. In the radical substitution mechanism this is done by using a dot to track the formation and reaction of the radicals. In (e) make sure you follow the instructions and use your data sheet.

OCR(A) AS Chemistry

Student A

(a) $Cl_2 \rightarrow 2Cl\bullet$

$C_2H_6 + Cl\bullet \rightarrow HCl + C_2H_5\bullet$

$C_2H_5\bullet + Cl_2 \rightarrow C_2H_5Cl + Cl\bullet$

$Cl\bullet + Cl\bullet \rightarrow Cl_2$

Student B

(a) $Cl_2 \rightarrow 2Cl\bullet$

$C_2H_6 + Cl\bullet \rightarrow H\bullet + C_2H_5Cl$

$H\bullet + Cl_2 \rightarrow HCl + Cl\bullet$

$C_2H_5\bullet + Cl\bullet \rightarrow C_2H_5Cl$

ⓔ Student A gains all 4 marks, while student B only picks up the first and last marks. The propagation steps shown by student B are incorrect. The initial propagation step of the Cl• with any alkane *always* produces an alkyl radical and HCl.

Student A

(b) **(i)** 1,2-dichloroethane

Student B

(b) **(i)** 1,2-dichloroethane

ⓔ Correct, for 1 mark.

Student A

(b) (ii)

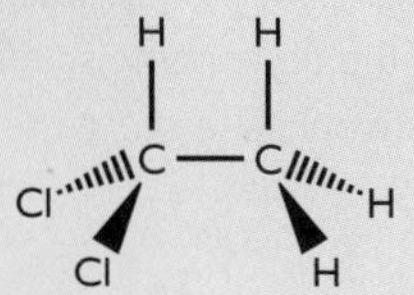

Student B

(b) (ii)

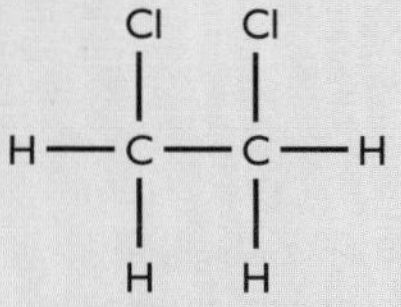

Z-isomer
(*cis*)

ⓔ Student A gains the mark, but student B has fallen into the trap set by the examiner. It is not possible to get *E/Z* isomerism (*cis–trans*) unless there is a C=C double bond. The isomer drawn by student B is identical to the one given in the question, because the atoms can rotate about the C–C single bond.

Student A

(c) **(i)** Water

Student B

(c) (i) (aq)

ⓔ Both answers score the mark.

Student A

(c) (ii) A nucleophile

Student B

(c) (ii) Donates a lone pair of electrons

ⓔ Each answer is worth 1 out of 2 marks. Neither student has used the information in the question. There are 2 marks available and therefore the examiner requires two separate points: *state* that it behaves as a nucleophile ✔ and *explain* that it is because it donates a lone pair of electrons ✔.

Student A

(d) (i) Heat with a condenser

Student B

(d) (i) Evaporates and then condenses

ⓔ Neither student scores the mark. To get the mark you must state that refluxing involves *continuous* evaporation and condensation, so that the volatile reagents/products are retained in the reaction vessel. Student B is almost correct, but has missed the key word *continuous*. Student A recalls that a condenser is needed, but is too vague to gain a mark. If the description is difficult to put into words, then you might want to draw a diagram instead. This would also score the marks.

Student A

(d) (ii) Orange to green

Student B

(d) (ii) Turns green

ⓔ Student A gains the mark, while student B again loses the mark by only giving the final colour. When recording observations, it is important to write down the initial observation as well as the final observation.

Student A

(d) (iii) $C_2H_5OH + [O] \rightarrow CH_3CO_2H + H_2O$

Student B

(d) (iii) $C_2H_6O + [O] \rightarrow C_2H_4O_2 + H_2$

ⓔ Student A gains 1 mark for the correct products, but loses a mark because the equation isn't balanced. When trying to balance equations students often forget the O in the OH of the alcohol.

Student B loses both marks and writes the most common incorrect answer. The formula $C_2H_4O_2$ is ambiguous and doesn't clearly identify the organic product as ethanoic acid. At first glance, it looks good because it is balanced, *but* H_2 is not a product. Water is *always* formed. The correct equation is $C_2H_5OH + 2[O] \rightarrow CH_3CO_2H + H_2O$.

Student A

(e) 2500–3300 cm^{-1}. It is an OH bond.
   1680–1750 cm^{-1}. It is a C=O bond.

Student B

(e) 3000 cm^{-1}. It is an OH bond.
   1700 cm^{-1}. It is a C=O bond.

ⓔ Both students gain all 4 marks, although student A again displays better examination technique. The absorptions are listed on the data sheet supplied in the exam. Student A has used the sheet and copied out the values, while student B has relied on memory.

ⓔ **Once again, student A scores more than student B (14 out of 17 compared with 9 out of 17). Student A scores 82% (a grade A), while student B scores 53% (grade D).**

Question 5 Electrophilic addition

Time allocation: 6–7 minutes

Describe and explain how propene reacts with bromine. You should include in your answer:
• **any observations**
• **the name of any organic products**
• **a full description of the mechanism**   (8 marks)

In this question, 1 mark is available for the quality of written communication.   (1 mark)

Total: 9 marks

ⓔ The command word 'describe' requires significant detail. When referring to a mechanism, equations are required and the movement of electrons must be tracked. In the electrophilic addition mechanism this is done by showing curly arrows, relevant dipoles and lone pairs of electrons. In a free-response question like this it is also useful to plan out how you are going to answer before you start.

Student A

When bromine reacts with propene it undergoes electrophilic addition. ✔ As the Br–Br molecule approaches the C=C double bond, it forms a temporary induced dipole ✔ in the Br_2, and the $Br^{\delta+}$ end is attracted to the alkene. The Br–Br bond is broken by heterolytic fission ✔ and results in a carbocation ✔. The Br^- then attacks the carbocation to form the product ✔. The product is called 1,2-dibromopropane. ✔ QWC ✔

Student B

e Use the following mark scheme to mark each of the student's answers. Check your marking against that of the examiner.

- observation: (bromine is) decolorised (do not accept 'clear') (1)
- product: 1,2-dibromopropane (1)
- electrophilic addition (1)
- induced dipole in the Br₂/dipoles shown correctly on the Br–Br bond and curly arrow on Br–Br bond as shown/heterolytic fission (1)
- curly arrow from the π-bond to the bromine, or words to that effect (1)
- intermediate carbonium ion/carbocation (1)
- curly arrow from Br⁻ back to the carbonium ion/carbocation/nucleophilic attack/Br⁻ forms a covalent bond with the carbocation (1)
- lone pair of electrons shown on the Br⁻ (and curly arrow from lone pair to the carbonium ion/ carbocation)/Br⁻ acts as a lone pair donor (1)

The mark for quality of written communication (QWC) is awarded for the description or layout of the mechanism, making use of appropriate chemical terms/symbols. Students often find free-response questions difficult to answer. In the question you are asked to state any observations and name any organic products and you are informed that 1 mark is available for the quality of written communication. This accounts for 3 of the available 9 marks, leaving 6 marks for the mechanism; therefore six separate points are required.

Both students have tackled this question well, but differently. Student A has written in prose and described the mechanism well. Describing a mechanism this way is difficult and requires a great deal of thought, planning and care. Student A clearly understands the mechanism, but has failed to show or to describe in words, the movement of electrons. He/she has concentrated so hard on describing the mechanism that the observation has been omitted. This is easily done.

Student B has taken the opposite approach — few or no words — and many students communicate like this. Student B has scored 5 of the 6 marks for the mechanism as well as the marks for stating the observation and naming the organic product. However, he/she has failed to state that the mechanism is an electrophilic addition mechanism, and hence loses 1 mark. In addition, student B has lost the mark for quality of written communication simply because there is no writing. Examiners accept that chemists communicate in a variety of ways: equations, diagrams and tables. None of these is penalised when it comes to awarding quality of written communication marks, *but* whenever there are marks for quality of written communication, students must write at least two consecutive sentences, ensuring that they are relevant to the question.

e **Both students score 7 out of 9 marks. Strangely, student A's good examination technique may have cost a mark. Student A has recognised that marks are awarded for quality of written communication and has attempted to write all of the answer** in continuous prose. This is extremely difficult, and a simple observation mark has been lost. Student B has simply ignored the need for continuous prose, but has scored well. It is important to try to strike the right balance.

Question 6 Fuels

Time allocation: 6–7 minutes

The fractions from crude oil can be processed further by cracking, reforming and isomerisation. Outline, with the aid of suitable examples and equations, each of these processes. Explain the industrial importance of each process. (7 marks)

In this question, I mark is available for the quality of written communication. (1 mark)

Total: 8 marks

ℯ The command word 'outline' implies that a brief description of the essential detail is required and 'explain' infers that a reason(s) is needed to support your statements.

Student A

Cracking is used to break down long-chain hydrocarbons into shorter-chain hydrocarbons. High temperature and a catalyst are used. ✔ The shorter-chain molecules are in higher demand.

$C_{10}H_{22} \rightarrow C_8H_{18} + C_2H_4$✔

Isomerisation is used to convert straight-chain hydrocarbons into branched-chain hydrocarbons. The branched-chain hydrocarbons are added to petrol and increase the octane rating. ✔

✗

Reforming is the conversion of chain hydrocarbons into ring hydrocarbons. Ring compounds are also added to petrol because they burn more efficiently. ✔

$C_6H_{14} \longrightarrow$ $+ H_2$ ✗

QWC ✔

Student B

Crude oil is a complex mixture of hydrocarbons. The mixture varies throughout the world. The oil found in the North Sea around the UK is rich in gasoline and low in residues. American oil is the opposite. The crude oil undergoes fractional distillation, which separates the mixture by the components' boiling points. Small-chain hydrocarbons have low boiling points, while longer-chain molecules have higher

boiling points. The difference in boiling points can be explained by the extent of van der Waals forces between the molecules.

When the crude oil is fractionally distilled, it is first vaporised by passing it through pipes at about 400°C. The vapour then goes into a large column about 100m high; it slowly rises up the column, passing over trays as it does so. There is a temperature gradient in the column — it is hot at the bottom and cools down as the vapour rises up. This causes different fractions to condense at different points in the column. The most volatile fractions condense at the top and the least volatile at the bottom.

The following fractions are all separated out by fractional distillation: gases, gasoline, naphtha, kerosene, diesel, oil and residue. The residue is used for waxes and bitumen.

The most useful fraction is the gasoline, which is used as petrol. The demand for the gasoline fraction outstrips the amount of gasoline produced by fractional distillation. Therefore longer-chain hydrocarbons are cracked into smaller-chain gasoline-type molecules. Cracking is carried out at high temperature, ✔ using a fluidised bed where the vaporised oil fraction and the zeolite catalysts behave as a fluid. When cracking takes place, alkenes are always produced.

$$C_{10}H_{22} \rightarrow C_8H_{18} + C_2H_4 ✔$$

Ethene, C_2H_4, has many uses. It can be reacted with steam in the presence of a phosphoric acid catalyst at about 300°C and 200atm pressure to produce ethanol. Ethanol is widely used as a solvent and as the starting material for the production of many other useful chemicals. Ethene can also be polymerised to produce poly(ethene). ✔

Reforming is another technique used by the petrochemical industry to change chain hydrocarbons into ring...   QWC ✔

⊝ Student A has demonstrated good examination technique, following the guidelines given in the question. There are three processes, each requiring an explanation, an equation and a statement about industrial importance. Student A has three separate paragraphs, three equations and a brief statement about industrial importance. The use of the products from cracking is too vague, and the equations given for isomerisation and reforming are incorrect. Skeletal formulae are a good way of showing isomerisation, but the equation must balance. Look carefully at students A's response and compare it to the correct response, shown below.

Student A's answer

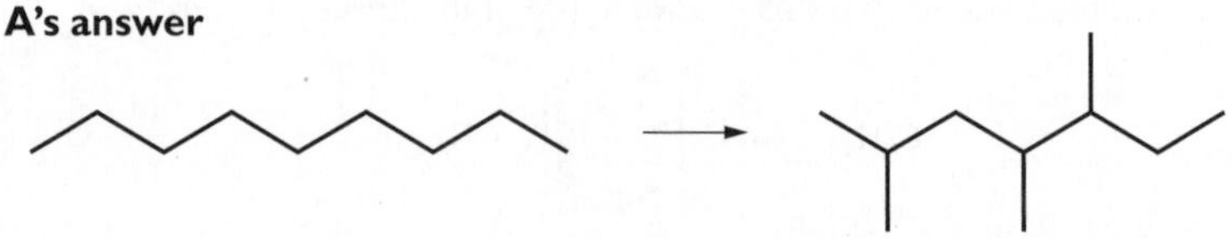

The reagent contains nine carbons (C_9H_{20}), but the product contains ten carbons, hence $C_{10}H_{22}$.

Examiner's answer

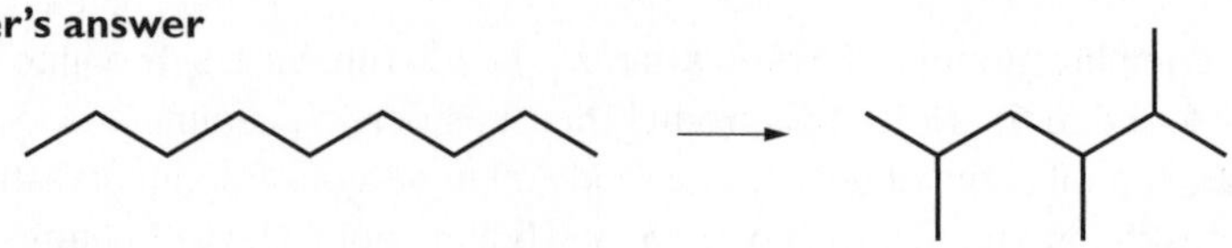

Both now contain nine carbons.

OCR(A) AS Chemistry

The equation for reforming isn't balanced, it should be either:

$$C_6H_{14} \longrightarrow \text{Cyclohexene} + 2 H_2$$

or

$$C_6H_{14} \longrightarrow \text{Cyclohexane} + H_2$$

Student B clearly is very knowledgeable about crude oil as a source of hydrocarbons — unfortunately, the question is about cracking, isomerisation and reforming. The opening three paragraphs display good knowledge and understanding, but relate to the fractional distillation of crude oil. In consequence, student B scores no marks for the opening comments. The paragraph relating to cracking also shows a deep understanding; student B scores 3 marks for this part of the question. It is clear from the fact that student B stopped 'mid-sentence' that time ran out. If student B had read the question carefully and planned the response, then there would have been enough time to complete the answer, which would probably have increased the overall mark.

The quality of written communication mark was awarded for spelling, punctuation and grammar, and both students were given this mark.

ⓔ Both students seem to understand this area of the specification, but their marks are low. Student A scores 5/8 and student B only 4/8. This reflects poor examination technique and a lack of precise, detailed knowledge. This is a common question, so you need to be well prepared to answer it.

Question 7 Boltzmann distribution

Time allocation: 6–7 minutes

The diagram below shows the energy distribution of reactant molecules at a temperature T_1.

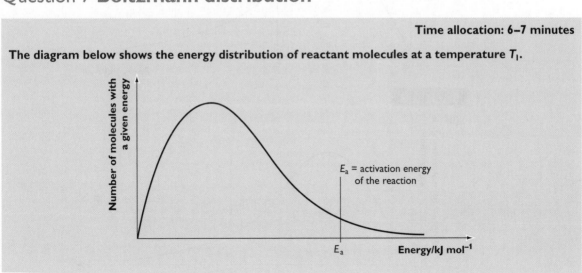

(a) Explain what is meant by the term activation energy. (1 mark)

(b) On the sketch above, mark E_c, the activation energy in the presence of a catalyst. (1 mark)

(c) Explain, in terms of the distribution curve, how a catalyst speeds up the rate of a reaction. (2 marks)

(d) (i) Raising the temperature can also increase the rate of this reaction. Draw a second curve to represent the energy distribution at a higher temperature. Label your curve T_2. (2 marks)
(ii) Explain how an increase in temperature can speed up the rate of a reaction. (2 marks)

Total: 8 marks

ⓔ The command word 'explain' in (a), (c) and (d) (ii) infers that reasons are needed to support your statements. The command word 'draw' in (d) (i) is self-explanatory and a diagram is required.

Student A

(a) The activation energy is the minimum energy needed to start a reaction.

Student B

(a) The energy needed for a collision to be successful.

ⓔ Student A gains the mark, but student B doesn't. The key word missing from student B's response is *minimum*.

Student A

(b)

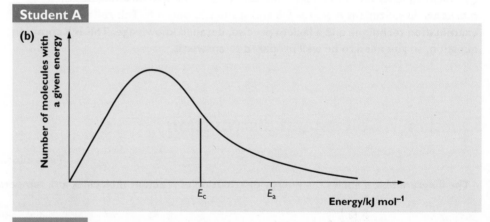

Student B

(b)

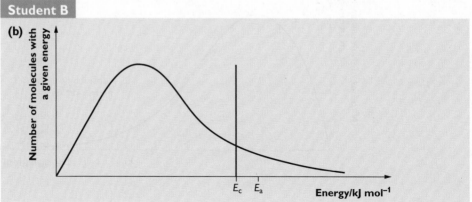

OCR(A) AS Chemistry

ⓔ Both score the mark. The activation energy for the catalyst must be lower than the original activation energy.

Student A

(c) The mode of action of any catalyst is to lower the activation energy, so that more particles now have enough energy to react.

Student B

(c) A catalyst is a substance that speeds up a reaction without itself being altered. Catalysts can be homogeneous (i.e. the same phase) or heterogeneous (a different phase). Catalysts can be reused.

ⓔ Student A scores both marks, but unfortunately student B gains no marks. Student B has not read the question carefully. He/she has simply written down correct information about catalysts, but nothing relevant to the question.

Student A

(d) (i)

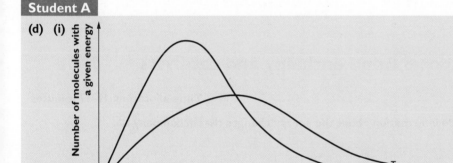

Student B

(d) (i)

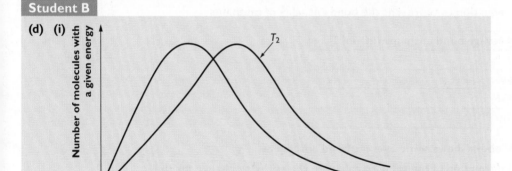

ⓔ Student A scores both marks, but student B drops a mark and only scores 1 out of the two possible marks. Both students show that at increased temperature the distribution moves to the right (to higher energy), and hence both score 1 mark. However, as the distribution moves to the right, the curve also flattens out, and hence student A scores the second mark but student B doesn't.

Student A

(d) (ii) Increasing temperature increases energy. Therefore more particles will have energy greater than or equal to the minimum energy required, and so the reaction will speed up because there will be more successful collisions.

Student B

(d) (ii) Increasing temperature lowers the activation energy, and hence more particles exceed the activation energy.

🅔 Student A gives the perfect response and gains both marks, but student B has misunderstood the effect that increasing temperature has on the activation energy, and has probably confused this effect with that of a catalyst. Unfortunately, student B scores no marks. (Catalysts lower E_a, but changing the temperature has no effect on the value of E_a.)

🅔 **The outcome is vastly different, with student A scoring full marks and student B only scoring 2 out of 8 marks (a grade U).**

Question 8 **Bond enthalpy and catalysts**

Time allocation: 11–12 minutes

Bond enthalpies can provide information about the energy changes that accompany a chemical reaction.

(a) What is meantby the term 'bond enthalpy'? (2 marks)

(b) (i) Write an equation, including state symbols, to represent the bond enthalpy of hydrogen chloride. (1 mark)
(ii) Write an equation to represent the bond enthalpy of methane. (2 marks)

(c)

Bond	Average bond enthalpy (kJ mol^{-1})
C–C	350
C=C	610
H–H	436
C–H	410

(i) The table above shows some average bond enthalpies.

Use the information in the table to calculate the enthalpy change for the hydrogenation of ethene: (3 marks)

Ethene (g)   Ethane (g)

> **(ii)** **The enthalpy change of this reaction can be found by experiment to be −136 kJ mol⁻¹.**
> **Explain why this value is different from that determined above.** (2 marks)
>
> **(d) In the above reaction, nickel is used as a catalyst.**
> **(i)** **Define 'catalyst'.** (1 mark)
> **(ii)** **Explain the mode of action of nickel in this reaction.** (3 marks)
>
> **Total: 14 marks**

ⓔ The command terms 'what is meant by' in (a) and 'define' in (d) (i) are in this context equivalent and only a simple statement is required. As in previous cases 'explain' requires reasons to justify your answers.

Student A

(a) It is the enthalpy change when 1 mole of a covalent bond is broken in the gaseous state.

Student B

(a) The energy needed to break 1 mole of a bond in the gas state.

ⓔ This term is difficult to define, but there are certain key elements that must be there, namely 'breaking a bond', and it must be in the 'gaseous state', and must involve '1 mole'. It is probably helpful to illustrate the definition with the equation: $X–Y(g) \rightarrow X(g) + Y(g)$, which shows that you also know that it relates to covalent molecules and involves homolytic fission. However, both students score both of the marks.

Student A

(b) **(i)** $HCl(g) \rightarrow H(g) + Cl(g)$

Student B

(b) **(i)** $HCl \rightarrow H + Cl$

ⓔ Student A scores the mark, but student B loses the mark, because the state symbols are essential.

Student A

(b) **(ii)** $CH_4(g) \rightarrow CH_3(g) + H(g)$

Student B

(b) **(ii)** $CH_4(g) \rightarrow C(g) + 4H(g)$

ⓔ Neither student scores both marks.

Student A has taken the definition of bond enthalpy literally and has 'broken 1 bond in the gaseous state', but because methane has four C–H bonds, the bond enthalpy is the average of all four bond enthalpies. Student A would be awarded 1 mark. Student B has shown the enthalpy change for breaking four C–H bonds, not one, and would also only score 1 mark.

The equation that the examiner is looking for is: $\frac{1}{4}[CH_4(g) \rightarrow C(g) + 4H(g)]$, which clearly indicates that the bond enthalpy in methane is the average bond enthalpy taken when all four C–H bonds are broken. This is shown below:

$$CH_4(g) \rightarrow CH_3(g) + H(g) \qquad \Delta H = +425\,kJ\,mol^{-1}$$
$$CH_3(g) \rightarrow CH_2(g) + H(g) \qquad \Delta H = +470\,kJ\,mol^{-1}$$
$$CH_2(g) \rightarrow CH(g) + H(g) \qquad \Delta H = +416\,kJ\,mol^{-1}$$
$$CH(g) \rightarrow C(g) + H(g) \qquad \Delta H = +335\,kJ\,mol^{-1}$$

The total enthalpy change is $(+425 + 470 + 416 + 335) = +1646\,kJ\,mol^{-1}$, which is the enthalpy change when four C–H bonds are broken in methane. Therefore, the enthalpy change when one C–H bond is broken is $+1646/4 = 411.5\,kJ\,mol^{-1}$. The students clearly are not expected to quote numerical values, but they are expected to realise that the bond enthalpy quoted is the average value.

Student A

(c) (i) Bonds broken $(610 + 1640 + 436) = +2686$
Bonds formed $(-350 - 2460) = -2810$
Enthalpy change $= -124\,kJ\,mol^{-1}$

Student B

(c) (i) Bonds broken: $1 \times$ C=C $\qquad\qquad +610$
Bonds formed: $2 \times$ C–H $\qquad\qquad -820$
$1 \times$ C–C $\qquad\qquad\qquad -350$
Enthalpy change: $\qquad\qquad -560\,kJ\,mol^{-1}$

ⓔ There are two ways of carrying out this calculation. Student A has opted for the safer way and calculated the enthalpy change for breaking every bond in the reactants and then for forming every bond in the reactant. A sensible way to set this out is to simply list each and every bond:

Bonds broken: $1 \times$ C=C $\qquad +610$
$\qquad\qquad\quad 4 \times$ C–H $\qquad +1640$
$\qquad\qquad\quad 1 \times$ H–H $\qquad +436$
Bonds formed: $1 \times$ C–C $\qquad -350$
$\qquad\qquad\quad 6 \times$ C–H $\qquad -2460$
Enthalpy change: $\qquad -124\,kJ\,mol^{-1}$

Student B has looked at the overall change and has attempted to identify the net change in bonds broken and formed. Student B has correctly worked out that if four C–H bonds are broken and six C–H bonds are formed, then the net change is the formation of two C–H bonds. However, student B has forgotten that H_2 consists of an H–H bond that also has to be broken. Student A scores all 3 marks, and student B gains 2 marks even though the answer is incorrect. This will be marked 'consequentially', as there is only one error in the calculation, and therefore student B will lose only 1 mark.

Student A

(c) (ii) The bond energies used in the calculation are average values for the bonds.

Student B

(c) (ii) Experiments are not very accurate and heat will be lost.

ⓔ Student A gains 1 mark and would have scored the second had he/she gone on to explain that the C–H bond in ethene will not be the same as the C–H bond in ethane, because they are in different environments. Student B has made unjustified assumptions about the accuracy of the experiment and scores no marks.

Student A

(d) (i) Speeds up a reaction without being used up.

Student B

(d) (i) Speeds up a reaction by lowering the activation energy.

ⓔ Both score the mark.

Student A

(d) (ii) The ethene and hydrogen gases are absorbed by the Ni, the reaction takes place, and the ethane is desorbed.

Student B

(d) (ii) The reactants bind to the surface of the Ni (adsorb) and the bonds are weakened. The reaction takes place and the product leaves the surface of the Ni (desorbs).

ⓔ The marking points are:
- adsorbs to Ni surface ✔
- weakens bonds/lowers activation energy ✔
- desorbs from Ni surface ✔

Student A only scores 1 out of 3 marks. By using the word 'absorbed', the first marking point is lost. In addition, no reference is made to how the bonds are weakened. Student B gains all 3 marks.

ⓔ **Student A scores 10 out of 14 marks, which is a B grade (71%). Student B scores only 1 mark less, but this makes it a C-grade answer (64%).**

Question 9 Enthalpy of combustion, Hess's law, catalytic converters

Time allocation: 9–10 minutes

Octane, C_8H_{18}, is one of the hydrocarbons present in petrol.

(a) Define the term 'standard enthalpy change of combustion'.   (3 marks)

(b) Use the data below to calculate the standard enthalpy change of combustion of octane:

$$C_8H_{18}(l) + 12\frac{1}{2}O_2(g) \rightarrow 8CO_2(g) + 9H_2O(l)$$

Compound	ΔH_f° (kJ mol^{-1})
$C_8H_{18}(l)$	−250.0
$CO_2(g)$	−393.5
$H_2O(l)$	−285.9

(3 marks)

(c) Combustion in a car engine also produces polluting gases, mainly carbon monoxide, unburnt hydrocarbons and oxides of nitrogen such as nitrogen(II) oxide, NO. Explain, with the aid of equations, how CO and NO are produced in a car engine. (2 marks)

(d) (i) The catalytic converter removes much of this pollution in a series of reactions. Write an equation showing the removal of carbon monoxide and nitrogen monoxide gases. (1 mark)

(ii) The removal of carbon monoxide and nitrogen monoxide gases involves a redox reaction. Use your answer to (d) (i) to identify the element being reduced and state the change in its oxidation number. (2 marks)

Total: 11 marks

ⓔ The command word 'define' in (a) requires only a simple statement. In (c), 'explain' requires only balanced equations as directed. In (d) (ii) 'identify' allows some flexibility: the name or formula is OK, but the identification must be unambiguous.

Student A

(a) It is the enthalpy change when 1 mole of a substance is burnt completely, in an excess of oxygen, under standard conditions of 298 K and 1 atmosphere.

Student B

(a) It's the enthalpy change when 1 mole of a substance is burnt in oxygen, under standard conditions.

ⓔ Student A gets all 3 marks. The marking points are:
- 1 mole ✔
- burnt in an excess of oxygen ✔
- standard conditions are 298 K/25°C and 100 kPa/1 atm ✔

Student B only scores 1 mark. Failing to specify the standard conditions is careless and demonstrates poor examination technique rather than a lack of knowledge.

Student A

(b)

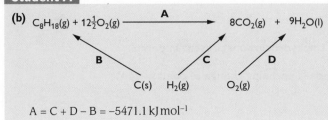

$A = C + D - B = -5471.1 \text{ kJ mol}^{-1}$

OCR(A) AS Chemistry

Student B

(b) $C_8H_{18}(g) + 12\frac{1}{2}O_2(g) \xrightarrow{\text{A}} 8CO_2(g) + 9H_2O(l)$

B
−250

C
8(−393.5)

D
9(285.9)

$C(s)$ $H_2(g)$ $O_2(g)$

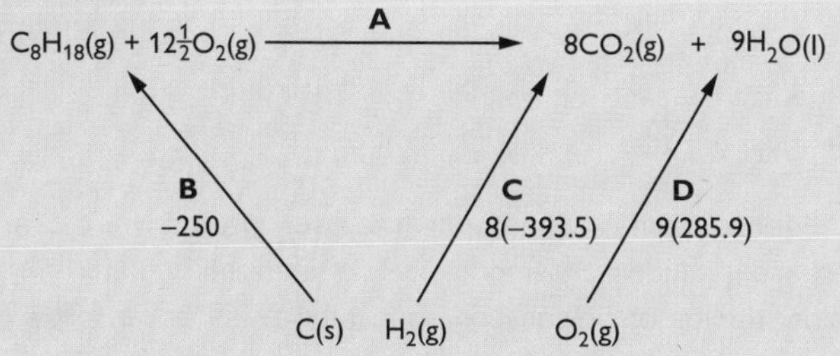

$A = C + D - B$

$= (-3148) + (2573.1) - (-250)$

$= -3148 + 2573.1 + 250$

$= -324.9\,kJ\,mol^{-1}$

ⓔ Student B shows better examination technique by giving all the working, but has forgotten to write the minus sign in front of the 285.9 for the formation of water. The examiner can clearly see this error and is able to track it through the working. Student B would therefore gain 2 of the 3 marks available. Although student A scores all 3 marks, had the numerical value been incorrect, he/she would have only scored 1 mark or possibly zero. *Remember:* it is always better to show all of your working in any calculation.

Student A

(c) $C_8H_{18}(l) + 8\frac{1}{2}O_2(g) \rightarrow 8CO(g) + 9H_2O(l)$
$N_2(g) + O_2(g) \rightarrow 2NO(g)$

Student B

(c) $C_8H_{18} + 9\frac{1}{2}O_2 + N_2 \rightarrow 8CO + 9H_2O + 2NO$

ⓔ Both students score 2 marks. Student B's response is a little unusual, but nevertheless correct.

Student A

(d) (i) $2NO(g) + 2CO(g) \rightarrow N_2(g) + 2CO_2(g)$

Student B

(d) (i) $2NO + 2CO \rightarrow N_2 + 2CO_2$

ⓔ Both students gain the mark. Marks for state symbols will only be awarded/deducted if they are asked for in the question.

Student A

(d) (ii) Nitrogen has been reduced because its oxidation state has changed from +2 to zero.

(d) (ii)

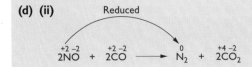

Reduced

$$\overset{+2\ -2}{2NO} + \overset{+2\ -2}{2CO} \longrightarrow \overset{0}{N_2} + \overset{+4\ -2}{2CO_2}$$

ⓔ Both students gain 2 marks. Again, student A has not shown any working, and would have lost both marks had the answer been wrong. Student B demonstrates better examination technique by writing the oxidation numbers along the top of the equation, so that the examiner can follow the working.

ⓔ **Both students have done well, and student A has scored full marks. However, student B carelessly lost 3 marks and scored 8 out of 11 marks (71%),which is a grade B.**

Question 10 **Equilibria**

Time allocation: 12–13 minutes

Sulfuric acid, H_2SO_4, is made industrially by the Contact process. This reaction is an example of a dynamic equilibrium:

$$2SO_2(g) + O_2(g) \rightleftharpoons 2SO_3(g) \qquad \Delta H = -98\,kJ\,mol^{-1}$$

(a) State two features of a reaction with a *dynamic equilibrium*. (2 marks)

(b) State and explain what happens to the equilibrium position of this reaction as:
(i) the temperature is raised (2 marks)
(ii) the pressure is increased (2 marks)
(iii) Suggest the optimum conditions for the Contact process. (2 marks)

(c) (i) The conditions used for the Contact process are a temperature of 450°C to 600°C and a pressure of around 10 atmospheres.
Explain why the optimum conditions are not used. (3 marks)
(ii) Vanadium(v) oxide is used as a catalyst. What effect does this have on the conversion of $SO_2(g)$ into $SO_3(g)$? (2 marks)
(iii) At least three catalyst chambers are used to ensure maximum conversion of $SO_2(g)$. The conversion yield can exceed 98%.
State two advantages of this high conversion rate. (2 marks)

Total: 15 marks

ⓔ The command word 'suggest' in (b) (iii) indicates that you are not expected to recall the values but you are meant to predict them using your answers from (b) (i) and (b) (ii).

(a) The rates of both reactions are the same, and therefore the amount of each chemical remains the same.

Student B

(a) The amount of each chemical in the system remains constant.
The reagents react at the same rate as the products.

ⓔ Both students gain both marks.

Student A

(b) (i) The position of the equilibrium moves to the left because the forward reaction is exothermic.

Student B

(b) (i) The reaction is much faster because more particles now exceed the activation energy.

ⓔ Student A scores both marks and has sensibly used the same words — 'equilibrium position' — as those used in the question. Unfortunately, student B has either misread or misunderstood the question, and has explained the effect of increasing temperature on the rate of reaction. Student B's explanation is correct, but it scores no marks because it does not address the question that was set. This is a common error, and many students lose valuable marks by not reading the question carefully.

Student A

(b) (ii) The equilibrium moves to the right because there are fewer molecules of gas on the right-hand side.

Student B

(b) (ii) Increasing pressure effectively increases the concentration, and therefore the reaction will go faster.

ⓔ Student B has again misread the question and made the same mistake as in (b) (i). This has cost another 2 marks.

Student A

(b) (iii) Low temperature and high pressure

Student B

(b) (iii) Low temperature and high pressure

ⓔ Both students gain 2 marks. This shows that student B clearly understands equilibrium and should have gained all 4 marks for (b) (i) and (b) (ii).

Student A

(c) (i) At low temperature the conversion is high, but the rate of reaction is too slow. High pressure is too expensive.

Student B

(c) (i) Temperature: a compromise is reached between rate and conversion.
At low temperature the rate is too slow.
Pressure: a compromise is reached between cost and conversion.
Catalyst: a catalyst is used to speed up the rate of conversion, so that it is cost effective to work at a low pressure.

e Student B gives the perfect answer and is awarded all 3 marks. Student B's response to this section demonstrates understanding and indicates that he/she should not have lost the marks in part (b). Student A just about scores 2 marks. The explanation of temperature is fine, but the explanation regarding pressure is barely adequate. As well as the use of a catalyst, the key to the answer is the compromise between the rate and percentage yield of SO_3 and the cost and percentage yield of SO_3.

Student A

(c) (ii) The catalyst speeds up the reaction, but it doesn't change the equilibrium position because it speeds up the forward and the reverse reactions equally.

Student B

(c) (ii) It speeds up the reaction by providing a mechanism of lower activation energy.

e Student A scores 2 marks, but again student B seems to have rushed the answer and has not fully addressed the question, scoring only 1 mark.

Student A

(c) (iii) More cost efficient and reduces the amount of SO_2 pollution

Student B

(c) (iii) More profitable and can make more money

e Student A gives a very good answer and gains both marks, but student B scores only 1 mark, having written the same thing twice.

e **Student B scores 9 marks for this question, which is just a grade C. However, if the student had not lost 4 marks in (b) (i) and (ii), then a grade A would have been awarded. Every mark is important, and it is essential to read the question carefully. Again, student A gives a good answer, scoring 14 out of 15 marks.**

Question 11 Activation-energy and energy-profile diagrams; le Chatelier's principle

Time allocation: 13–15 minutes

Hydrogen can be reacted with nitrogen in the presence of a catalyst to make ammonia, using the Haber process.

$$N_2(g) + 3H_2(g) \rightleftharpoons 2NH_3(g)$$

The activation energy for the forward reaction is $+68\,kJ\,mol^{-1}$.
The activation energy for the reverse reaction is $+160\,kJ\,mol^{-1}$.

(a) (i) Use this information to sketch the energy-profile diagram. Label clearly the activation energy for the forward reaction, E_f, and the activation energy for the reverse reaction, E_r. (2 marks)

(ii) Explain what is meant by 'activation energy'. (1 mark)

(iii) Calculate the enthalpy change for the forward reaction. (1 mark)

Much of the ammonia produced is oxidised into nitric acid, using the Ostwald process, which involves three stages:

Stage 1	$4NH_3(g) + 5O_2(g) \rightleftharpoons 4NO(g) + 6H_2O(g)$	$\Delta H = -950\,kJ\,mol^{-1}$
Stage 2	$2NO(g) + O_2(g) \rightleftharpoons 2NO_2(g)$	$\Delta H = -114\,kJ\,mol^{-1}$
Stage 3	$3NO_2(g) + H_2O(g) \rightleftharpoons 2HNO_3(g) + NO(g)$	$\Delta H = -117\,kJ\,mol^{-1}$

(b) (i) With reference to the oxidation number of the nitrogen in $NH_3(g)$ (stage 1) and HNO_3 (stage 3), show clearly that this is an oxidation process. (3 marks)

(ii) State le Chatelier's principle. (2 marks)

(iii) In stage 1, use le Chatelier's principle to predict and explain the temperature and the pressure that would give the maximum yield at equilibrium. (4 marks)

(iv) Suggest what happens to the NO(g) produced in stage 3. (1 mark)

(c) The nitric acid produced in stage 3 is a strong acid. Explain, with the aid of an equation, what is meant by the term strong acid. (2 marks)

Total: 16 marks

ⓔ The command word 'sketch' in (a) (i) clearly requires a diagram, but it is essential to ensure that the diagram is labelled, including any units.

Student A

(a) (i)

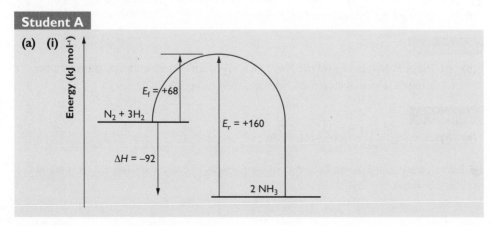

Student B

(a) **(i)**

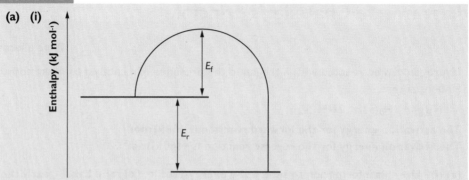

e The marking points are:
- products at a lower enthalpy than the reactants ✔
- activation energy of both forward and reverse reactions correctly labelled ✔

Student A gives the perfect answer and even goes on to work out ΔH for (a) (iii). Student B scores no marks. The activation energy for the reverse reaction is incorrect. Moreover, the energy levels are not labelled, so it is not possible to award the first marking point.

Student A

(a) **(ii)** Activation energy is the minimum energy required to start a reaction.

Student B

(a) **(ii)** It's the minimum energy needed to start the reaction.

e Both students gain the mark. The key phrase is the *minimum* energy.

Student A

(a) (iii) $-92\,\text{kJ}\,\text{mol}^{-1}$

Student B

(a) (iii) 92

e Student A gains the mark, but student B doesn't score anything because the sign of the enthalpy change for the forward reaction is negative.

Student A

(b) **(i)** The N in NH_3 is -3 and the N in HNO_3 is $+5$. An increase in oxidation number involves a loss of electrons, which is oxidation (OILRIG).

Student B

(b) **(i)** -3 to $+5$ $N^{-3} \rightarrow N^{+5} + 8e^-$

e Both students gain 3 marks. Each student has correctly deduced the oxidation numbers, and has stated or shown the loss of electrons.

Student A

(b) (ii) Le Chatelier's principle states that if a closed system under equilibrium is subject to a change, then the system will move in such a, way as to minimise the effect of the change.

Student B

(b) (ii) If concentration, temperature and pressure are changed, the system will move to oppose the change. If we increase the concentration, the system will move in such a way as to decrease the concentration. If we increase the temperature, the system will move in such a way as to decrease the temperature. If we decrease the pressure, the system will try to move in such a way as to increase the pressure.

e Both students gain 2 marks. Student A gives a textbook answer, while student B fully explains le Chatelier's principle.

Student A

(b) (iii) The forward reaction is exothermic and so will be favoured by low temperature. There are fewer products than reactants, so high pressure is required.

Student B

(b) (iii) Low temperature and high pressure, but in industry a temperature of about 450°C and a pressure of 200 atm are used.

e Student A scores 2 out of 4 marks. The prediction and explanation about temperature are correct and earn 2 marks. Student A concludes incorrectly that there are fewer products than reactants (the reverse is true), thus losing both marks relating to the required pressure. Student B gains 2 marks for the correct conditions, but doesn't score any marks for explaining why a low temperature is needed. Student B has simply memorised the industrial conditions.

Student A

(b) (iv) Reused in stage 2

Student B

(b) (iv) Recycled

e Both students gain the mark.

Student A

(c) Strong acids dissociate totally, e.g. $HCl \rightarrow H^+ + Cl^-$

Student B

(c) All acids have a pH below 7. Strong acids have a low pH of about 1 or 2.

e This question is worth 2 easy marks, but neither student scores both marks. You need to explain both 'strong' and 'acid'. Student A states that strong acids dissociate totally for 1 mark but

does not explain that acids are proton donors. If the student had written $HCl + H_2O \rightarrow H_3O^+$ + Cl^-, he/she would have scored both marks. Student B misses the point and gives a property of acids without addressing the question. He/she would probably score no marks.

ⓔ **Student A scores 12 out of 16 marks (75% — a grade B answer), while student B gains 9 out of 16 (56%, a grade D).**

Question 12 Enthalpy changes using $mc\Delta T$; catalysts

Time allocation: 14–15 minutes

In an investigation to find the enthalpy change of combustion of ethanol, C_2H_5OH, a student found that 1.60 g of ethanol could heat 150 g of water from 22.0°C to 71.0°C. The specific heat capacity of the apparatus is $4.2 \, J \, g^{-1} \, K^{-1}$.

(a) Use the results to calculate a value for the enthalpy change of combustion of ethanol. **(6 marks)**

(b) The theoretical value of the standard enthalpy change of combustion of ethanol is $-1367.3 \, kJ \, mol^{-1}$. Give two reasons for the difference in values, and suggest an improvement that could be made to the experiment, to minimise the most significant error. **(3 marks)**

(c) Catalysts have great economic importance. Give an example of a catalyst that is in the same state as the reactants, and an example of one that is in a different state to the reactants. State why the reaction you have chosen is important. Explain how the catalyst is able to increase the rate of reaction.
In this question, 1 mark is available for the quality of written communication. **(6 marks)**

Total: 15 marks

ⓔ The command word 'calculate' in (a) obviously requires a calculation *but* the calculation is worth 6 marks and you must show all of your working so that any errors can be tracked by the examiner. If a mistake is made the rest of the calculation can be marked consequentially.

Student A

(a) $Q = -mc\Delta T$
$= -(150 \times 4.2 \times 49)$
$= -30870 \, J = -30.87 \, kJ$
moles of ethanol = 1.6/46.0 = 0.035
$\Delta H = -30.87/0.035$
$= -882.8 \, kJ \, mol^{-1}$

Student B

(a) $Q = -mc\Delta T = -(1.6 \times 4.2 \times 49) = -329.28 \, J$
moles of ethanol = 1.6/46.0 = 0.035
$\Delta H = -329.28/0.035$
$= -9408$

ⓔ The marking points are:
- use the equation $Q = -mc\Delta T$ ✔
- mass = 150 ✔
- calculate the value of Q ✔
- moles of ethanol ✔
- divide the value of Q by moles of ethanol ✔
- use the correct units ✔

Student A's answer is fine, even though the correct answer is $887\,kJ\,mol^{-1}$. Student A doesn't obtain this answer, because he/she has rounded the number of moles of ethanol to 0.035 and has then used this figure in the next step of the calculation. This is an incorrect procedure, as the full number should be held in the calculator, to be used in subsequent steps. However, student A would probably be awarded either all 6 marks or at least 5 out of 6. Student B uses the incorrect value for the mass, and doesn't quote any units for the final value, but still scores 4 marks, even though his/her answer is wrong. Student B showed all the working, and the examiner was able to award marks consequentially following the initial mistake in the calculation.

Student A

(b) Heat is lost to the surroundings.
The thermometer wasn't very accurate.
I would use a more accurate thermometer to reduce the error.

Student B

(b) Heat loss is the most significant error.
Incomplete combustion. CO or C may have been formed.
Increase the supply of oxygen to ensure that CO_2 is always produced.

ⓔ Student B gives a good answer and scores 2 marks, having identified the most significant error as heat loss, but hasn't suggested how to modify this error. Student A also gains 2 marks. Students often criticise the apparatus without carefully considering the error. Given that the temperature rise was approximately 50°C, if the thermometer was able to measure to the nearest degree, then the percentage error in measuring the temperature rise is of the order $(1/50) \times 100 = 2\%$. Given the inaccuracy of the experiment, this is not the most significant error.

Student A

(c) Catalysts speed up the rate of reactions and therefore reduce costs. Sulfuric acid is used in the production of esters. ✔ The acid is in the same state as the reactants and is known as a homogeneous catalyst. Heterogeneous catalysts, such as Fe, which is used in the Haber process, are catalysts that are in a different state to the reactants. ✔ Both types of catalyst work by lowering the activation energy for the reaction. ✔   QWC ✔

Student B

(c) Esters are made from the reaction between alcohols and carboxylic acids in the presence of an acid ✔ catalyst. The acid and the reactants are all liquids, and the catalyst is a homolytic ✗ catalyst. Esters are important, as they are used in flavourings ✔. Platinum ✔ is used in the catalytic converter to reduce the amount of CO and NO ✔ emitted. The platinum is a solid and the reactants are both gases.

This is a heterolytic ✗ catalyst. Catalysts speed up a reaction, without being used up, by providing an alternative route of lower activation energy. ✔

🅔 The marking points are:
- identification of a suitable example of a homogeneous catalyst ✔
- suitable use for product ✔
- identification of a suitable example of a heterogeneous catalyst ✔
- suitable use for product ✔
- catalysts lower the activation energy ✔

Quality of written communication marks are awarded for: appropriate spelling, punctuation, grammar and correct use of at least two of the following: 'homogeneous', 'heterogeneous' and 'activation energy'. ✔

Student A gives a very good answer, but forgets to state a use for each process and therefore loses 2 marks. Student A scores 1 mark for each example; 1 mark for the explanation of how the catalyst works; and gains 1 mark for 'quality of written communication'. Student B scores 5 marks. Student B has given an example and a use of each type, and has correctly explained how catalysts work. The only mark lost is for 'quality of written communication', because student B has confused homolytic and heterolytic (which relates to bond fission) with homogeneous and heterogeneous.

🅔 **Each student scores 11 out of 15 marks — a grade B. Student A could have earned a grade A by checking that his/her answer to part (c) addressed all aspects of the question. Student B should have recognised that the *m* in *mcΔT* refers to total mass and not to an individual reagent.**

Question 13 The greenhouse effect; removal of carbon dioxide

Time allocation: 8–9 minutes

Many manufacturing processes, such as the production of calcium oxide from limestone, produce carbon dioxide as an unwanted by-product. Carbon dioxide contributes to global warming.

(a) Using carbon dioxide as your example, outline how global warming occurs. (3 marks)

(b) Explain one possible way that waste carbon dioxide could be removed. (2 marks)

(c) Treatment of waste products is sometimes necessary for industrial processes, but it is usually expensive. It is better to design processes that are more environmentally friendly. Suggest three features of the design of a manufacturing process that are desirable in order to minimise damage to the environment. (3 marks)

Total: 8 marks

🅔 The command word 'outline' in (a) requires a brief response but it is always worth looking at the mark allocation: 3 marks requires three separate points. In (c), 'suggest' requires application of knowledge extended beyond the remit of the specification.

OCR(A) AS Chemistry

Student A

(a) Carbon dioxide is able to absorb infrared radiation and this doesn't allow it to escape into outer space. The infrared radiation then causes the atmosphere to heat up.

Student B

(a) Carbon dioxide acts as a blanket above the atmosphere, and it therefore traps any escaping heat and keeps the Earth warm. It does it by vibrating its bonds.

ⓔ Both these answers have some general merit, but in neither case has the student appreciated the detail that is expected.

The marking scheme is as follows:
- The C=O bond in carbon dioxide absorbs infrared radiation ✔ by causing the bond to vibrate. ✔
- Some infrared radiation is trapped and causes increased heating of the atmosphere. ✔

On this basis, student A gains 1 mark for mentioning the heating of the atmosphere. Student B might just be allowed 1 mark, but is more likely to score zero.

When revising the environmental sections of the specification it is essential to make sure that specific points have been learnt and that the subject is not treated in an oversimplistic way.

Student A

(b) Carbon dioxide can be removed using carbon capture and storage. The gas is liquefied and then injected into rocks deep beneath the oceans.

Student B

(b) The carbon dioxide can be absorbed into an alkali.

ⓔ Both answers give a correct method. Student A provides sufficient detail for the 2 marks. However, student B's explanation is superficial, as it provides no detail. The answer is only worth 1 mark; writing a suitable balanced equation would have secured the second mark.

Student A

(c) The three features are:
- Use a catalyst where possible.
- Use renewable materials for the process.
- Minimise the formation of unwanted waste products.

Student B

(c) Don't design processes which require lots of energy.
Make sure the product you are making doesn't have side-effects.
Don't dump waste materials.

ⓔ Once again, student A has taken more trouble to study the appropriate section of the specification. The answer is worth 2 of the 3 marks. The use of a catalyst in this context should be justified in terms of the likelihood that it would lead to a lower energy requirement, rather than it

speeding up the reaction. Therefore, without a supporting comment, the first statement of student A's answer is insufficient for a mark.

Student B understands the point about low-energy requirements, and therefore scores 1 mark. The other two points that were made raise sensible issues, but they do not answer the question, which referred to the *design* of the process.

ⓔ This section of the specification is perhaps the most daunting, as it deals with current environmental issues. When answering questions about global warming, you must relate the issues to chemical facts and principles. Student A does this better than student B. Student A scores 5 out of 8, while student B scores only 2 out of 8.

OCR(A) AS Chemistry

Knowledge check answers

1 2-bromopropane, but-2-ene, 2,2-dimethylpropane or dimethylpropane, 3-methylpentane

2

H—C—C—C—H (with Cl, H groups) **2-chloropropane**

H—C—C—C—C—H (with OH group) **butan-2-ol**

H—C—C—C=C (3-methylbut-1-ene) **3-methylbut-1-ene**

3

pent-2-ene   **butan-2-ol** (OH)

3-chlorobut-1-ene (Cl)

4 The equation $C_2H_5OH + CH_3COOH \rightarrow CH_3COOC_2H_5 + H_2O$ shows mole ratios are 1:1:1:1.
Moles of C_2H_5OH (5.00/46.0 = 0.10869562)*; moles of CH_3COOH (8.00/60.0 = 0.13333)*, which shows that CH_3COOH is in excess and the maximum number of mols of ester that could be produced is 0.11. Number of moles of ester actually produced is 7.12/88.0 = 0.0809090* and the % yield is 74.4%. (* Do not round during the calculation.)

5 2,3-dimethylbutane, 2-methylpentane, hexane, octane

6 $C_{12}H_{26} \rightarrow C_9H_{20} + C_3H_6$ $C_{12}H_{26} \rightarrow C_7H_{16} + C_3H_6 + C_2H_4$

7 $C_3H_6 + 4\frac{1}{2}O_2 \rightarrow 3CO_2 + 3H_2O$

8 $H_2C=CHCH=CH_2 + 2Br_2 \rightarrow CH_2BrCHBrCHBrCH_2Br$; the Br_2 would be decolorised. Organic product: 1,2,3,4-tetrabromobutane.

9 $CH_3CH=CHCH_3 + H_2O \rightarrow CH_3CH_2CH(OH)CH_3$

10 An electrophile is an electron pair acceptor.
$C_6H_{10} + Br_2 \rightarrow C_6H_{10}Br_2$
or

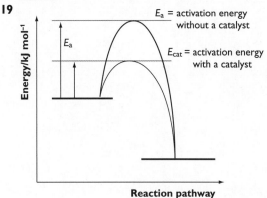

11

$nCH_3CHCHCH_3 \longrightarrow$ (polymer structure with CH_3 CH_3 groups, repeating unit n)

(structure: CH_3 CH_3 CH_3 CH_3 chain)

12 $C_3H_7OH + 4\frac{1}{2}O_2 \rightarrow 3CO_2 + 4H_2O$

13 $CH_3CH_2CH(OH)CH_2CH_3 \rightarrow CH_3CHCHCH_2CH_3 + H_2O$

14 $CH_3CH(OH)CH_2CH_3 + [O] \rightarrow CH_3COCH_2CH_3 + H_2O$

15 Reflux: continuous evaporation and condensation such that volatile components cannot escape.
Distillation: evaporation followed by condensation such that volatile components can escape.

16 The C—Cl bond is longer and weaker than the C—F bond.

17 CH_3^+, $CH_3CH(OH)^+$, $CH_3CH(OH)CH_2^+$, $CH_3CH(OH)CH_2CH_3^+$

18 Activation energy is the minimum energy required to start a reaction.

19

Energy/kJ mol⁻¹ vs Reaction pathway graph:
E_a = activation energy without a catalyst
E_{cat} = activation energy with a catalyst

20 The enthalpy change when 1 mole of substance is formed from its elements under standard conditions of 298 K and 101 kPa.
$3C(s) + 3H_2(g) + \frac{1}{2}O_2(g) \rightarrow CH_3COCH_3(l)$

21 The enthalpy change when 1 mole of substance is burnt in excess oxygen under standard conditions of 298 K and 101 kPa.
$CH_3CH_2CHO(l) + 4O_2(g) \rightarrow 3CO_2(g) + 3H_2O$

22 Increasing pressure forces the gaseous particles closer together (volume decreases) and increases the chance of a collision.

23 A reversible reaction is one whereby the reagents react to form the products but also the products react to re-form the reagents.
A dynamic equilibrium is when the rate of the forward reaction equals the rate of the reverse reaction, such that the concentration of the reagents and products remains constant, while the reagents and the products constantly interchange.

24 Le Chatelier's principle states that if a closed system under equilibrium is subject to a change, then the system will move in such a way as to *minimise* the effect of the change.

25 When a gas absorbs infrared it causes the bonds to vibrate and/or stretch and/or bend and prevents the infrared from escaping. The infrared is released by the greenhouse gas and most of it is returned to the Earth's surface.

26 Its ability to absorb infrared radiation

Its atmospheric concentration

Its residence time — how long it stays in the atmosphere

27 By converting it into liquid CO_2 it can then be injected deep underground or under the oceans for storage. Possible sites include depleted oil and gas fields.

By reacting the carbon dioxide with metal oxides to form carbonates.

28 Ozone, O_3, is being formed continuously and broken down in the stratosphere, so that an equilibrium is established with molecular oxygen, O_2, and an oxygen radical, O.

$$O_2 + O \rightleftharpoons O_3$$

29 $CCl_2F_2 \rightarrow \cdot CClF_2 + Cl\cdot$

30 $NO + O_3 \rightleftharpoons NO_2 + O_2$

$NO_2 + O \rightleftharpoons NO + O_2$

The net reaction is $O_3 + O \rightarrow 2O_2$ with the NO behaving as a catalyst.

31 Removal of carbon monoxide:

$2NO(g) + 2CO(g) \rightarrow N_2(g) + 2CO_2(g)$

and oxides of nitrogen:

$2NO_2(g) + 4CO(g) \rightarrow N_2(g) + 4CO_2(g)$

32 $N_2(g) + O_2(g) \rightarrow 2NO(g)$